Biology: The Science for Life

Biology: The Science for Life

Anthony Perez

STATES
ACADEMIC PRESS
www.statesacademicpress.com

States Academic Press,
109 South 5th Street,
Brooklyn, NY 11249, USA

Visit us on the World Wide Web at:
www.statesacademicpress.com

ISBN: 978-1-63989-078-1 (Hardback)

Cataloging-in-Publication Data

Biology : the science for life / Anthony Perez.
p. cm.
Includes bibliographical references and index.
ISBN 978-1-63989-078-1
1. Biology. 2. Life sciences. 3. Natural history. 4. Life (Biology).
I. Perez, Anthony.
QH307.2 .B56 2022
570--dc23

Table of Contents

Preface

Biology is a branch of science which is concerned with the study of life. It focuses on the physical structure, physiological mechanisms, development and evolution of living organisms. Cell is considered to be the basic unit of life within this field. The discipline of biology is broadly classified into theoretical biology and experimental biology. Theoretical biology makes use of mathematical techniques for the formulation of quantitative models. Empirical biology deals with experiments which are used for testing the validity of proposed theories. Some of the basic concepts of modern biology are cell theory, evolution, genetics and homeostasis. Certain biological phenomena such as cell division, reproduction and transmission of genetic material are also studied in this field. The topics included in this book on biology are of utmost significance and bound to provide incredible insights to readers. With its detailed analyses and data, it will prove immensely beneficial to professionals and students involved in this area at various levels.

A detailed account of the significant topics covered in this book is provided below:

Chapter 1- Biology is the study of life, dealing with the structure, function, growth, origin, evolution and distribution of living organisms. It plays an important role in understanding the living world. This chapter introduces biology and its various concepts to the reader for a better understanding.

Chapter 2- Cells are the fundamental units of life. The ability of cells to divide led to the creation of advanced multicellular life forms. Contained within these cells are genes, through which traits are passed down from one generation to the next. This chapter seeks to explain the topic in a simple and detailed manner for the reader's benefit.

Chapter 3- Human biology deals with human beings and population. All aspects of human functionality, be it genetics, anatomy, physiology, anthropology, etc. are studied under human biology. It is an interdisciplinary study with close ties to biomedical sciences, biomedical anthropology and many other studies related to the human body. This chapter provides a detailed explanation of the mentioned fields.

Chapter 4- Botany is the scientific study of plants, their structure, growth, development, classification, diseases and environmental impacts. Considering the fundamental role of plants in human survival, the importance of botany has increased significantly. Plant anatomy, breeding, propagation, morphology, ecology, physiology and pathology are the important topics covered in this chapter.

Chapter 5- The study of any life form that is mobile and exhibits a heterotrophic mode of nutrition comes under animal biology. Animals come in different shapes and sizes but display similar characteristics at the cellular level. This chapter covers in detail all the varied topics that come under it, like animal physiology, animal anatomy and animal behavior.

Chapter 6- All living organisms procreate to maintain a continuity of their species. This biological process by which an organism replicates itself is known as reproduction and it can be either asexual or sexual. This chapter provides an extensive explanation of the various biological processes like spore formation, fertilization, fusion, fission, etc. which can be found in plant, animal and human reproduction.

It gives me an immense pleasure to thank our entire team for their efforts. Finally in the end, I would like to thank my family and colleagues who have been a great source of inspiration and support.

Anthony Perez

Chapter 1

Understanding Biology

Biology is the study of life, dealing with the structure, function, growth, origin, evolution and distribution of living organisms. It plays an important role in understanding the living world. This chapter introduces biology and its various concepts to the reader for a better understanding.

Biology is the study of everything that is, or was once, alive — whether it's a plant, animal or micro-organism. Biology is the study of life. The word "biology" is derived from the Greek words "bios" (meaning life) and "logos" (meaning "study"). In general, biologists study the structure, function, growth, origin, evolution and distribution of living organisms. Biology is important because it helps us understand how living things work and how they function and interact on multiple levels. Advances in biology have helped scientists do things such as develop better medicines and treatments for diseases, understand how a changing environment might affect plants and animals, produce enough food for a growing human population and predict how eating new food or sticking to an exercise regimen might affect our bodies.

Biology is the study of all living organisms, including this adorable jewel bug and the plant it's perched on.

The Basic Principles of Biology

Four principles unify modern biology:

- Cell theory is the principle that all living things are made of fundamental units called cells, and all cells come from preexisting cells.
- Gene theory is the principle that all living things have DNA, molecules that code the structures and functions of cells and get passed to offspring.
- Homeostasis is the principle that all living things maintain a state of balance that enables organisms to survive in their environment.

- Evolution is the principle that describes how all living things can change to have traits that enable them to survive better in their environments. These traits result from random mutations in the organism's genes that are "selected" via a process called natural selection. During natural selection, organisms that have traits better-suited for their environment have higher rates of survival, and then pass those traits to their offspring.

Properties of Life

All groups of living organisms share several key characteristics or functions: order, sensitivity or response to stimuli, reproduction, adaptation, growth and development, regulation/homeostasis, and energy processing. When viewed together, these eight characteristics serve to define life.

Order

Organisms are highly organized structures that consist of one or more cells. Even very simple, single-celled organisms are remarkably complex. Inside each cell, atoms make up molecules. These in turn make up cell components or organelles. Multicellular organisms, which may consist of millions of individual cells, have an advantage over single-celled organisms in that their cells can be specialized to perform specific functions, and even sacrificed in certain situations for the good of the organism as a whole.

A toad represents a highly organized structure consisting of cells, tissues, organs, and organ systems.

Sensitivity or Response to Stimuli

Organisms respond to diverse stimuli. For example, plants can bend toward a source of light or respond to touch as shown in figure below. Even tiny bacteria can move toward or away from chemicals (a process called chemotaxis) or light (phototaxis). Movement toward a stimulus is considered a positive response, while movement away from a stimulus is considered a negative response.

The leaves of this sensitive plant (Mimosa pudica) will instantly droop and fold when touched. After a few minutes, the plant returns to its normal state.

Reproduction

Single-celled organisms reproduce by first duplicating their DNA, which is the genetic material, and then dividing it equally as the cell prepares to divide to form two new cells. Many multicellular organisms (those made up of more than one cell) produce specialized reproductive cells that will form new individuals. When reproduction occurs, DNA containing genes is passed along to an organism's offspring. These genes are the reason that the offspring will belong to the same species and will have characteristics similar to the parent, such as fur color and blood type.

Adaptation

All living organisms exhibit a "fit" to their environment. Biologists refer to this fit as adaptation and it is a consequence of evolution by natural selection, which operates in every lineage of reproducing organisms. Examples of adaptations are diverse and unique, from heat-resistant Archaea that live in boiling hot springs to the tongue length of a nectar-feeding moth that matches the size of the flower from which it feeds. All adaptations enhance the reproductive potential of the individual exhibiting them, including their ability to survive to reproduce. Adaptations are not constant. As an environment changes, natural selection causes the characteristics of the individuals in a population to track those changes.

Growth and Development

Organisms grow and develop according to specific instructions coded for by their genes. These genes provide instructions that will direct cellular growth and development, ensuring that a species' young as shown in figure below will grow up to exhibit many of the same characteristics as its parents.

Although no two look alike, these kittens have inherited genes from both parents and share many of the same characteristics.

Regulation/Homeostasis

Even the smallest organisms are complex and require multiple regulatory mechanisms to coordinate internal functions, such as the transport of nutrients, response to stimuli, and coping with environmental stresses. For example, organ systems such as the digestive or circulatory systems perform specific functions like carrying oxygen throughout the body, removing wastes, delivering nutrients to every cell, and cooling the body.

To function properly, cells require appropriate conditions such as proper temperature, pH, and concentrations of diverse chemicals. These conditions may, however, change from one moment to the next. Organisms are able to maintain internal conditions within a narrow range almost constantly, despite environmental changes, through a process called homeostasis or "steady state"—the ability of an organism to maintain constant internal conditions. For example, many organisms regulate their body temperature in a process known as thermoregulation. Organisms that live in cold climates, such as the polar bear in figure below, have body structures that help them withstand low temperatures and conserve body heat. In hot climates, organisms have methods (such as perspiration in humans or panting in dogs) that help them to shed excess body heat.

Polar bears and other mammals living in ice-covered regions maintain their body temperature by generating heat and reducing heat loss through thick fur and a dense layer of fat under their skin.

Energy Processing

All organisms use a source of energy for their metabolic activities. Some organisms capture energy from the Sun and convert it into chemical energy in food; others use chemical energy from molecules they take in. A lot of energy is required for a California condor to fly. Chemical energy derived from food is used to power flight. California condors are an endangered species; scientists have strived to place a wing tag on each bird to help them identify and locate each individual bird.

Evolution

The diversity of life on Earth is a result of mutations, or random changes in hereditary material over time. These mutations allow the possibility for organisms to adapt to a changing environment. An organism that evolves characteristics fit for the environment will have greater reproductive success, subject to the forces of natural selection.

Levels of Organization of Living Things

Living things are highly organized and structured, following a hierarchy on a scale from small to large. The atom is the smallest and most fundamental unit of matter. It consists of a nucleus surrounded by electrons. Atoms form molecules. A molecule is a chemical structure consisting of at least two atoms held together by a chemical bond. Many molecules that are biologically important are macromolecules, large molecules that are typically formed by combining smaller units called monomers. An example of a macromolecule is deoxyribonucleic acid (DNA) is shown below, which contains the instructions for the functioning of the organism that contains it.

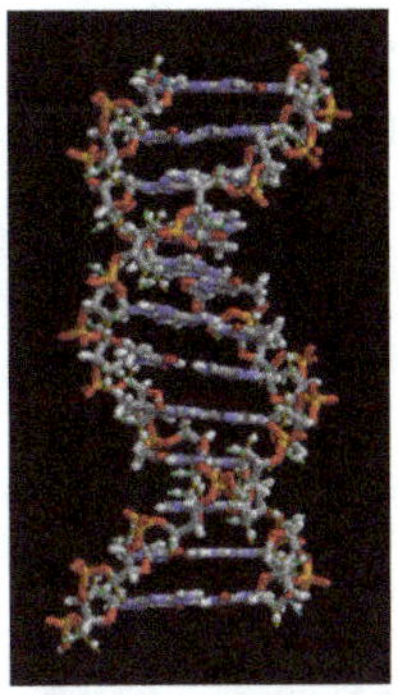

A molecule, like this large DNA molecule.

Kingdoms of Life

The earliest classification system recognized only two kingdoms: plants and animals but the use of microscope led to discovery of microorganisms, so the two kingdoms system was no longer useful. Today the system of classification includes six kingdoms in which two kingdoms of prokaryotes and four kingdoms of eukaryotes. The organisms are classified in to three domains: Bacteria, Archea and Eukaryota.

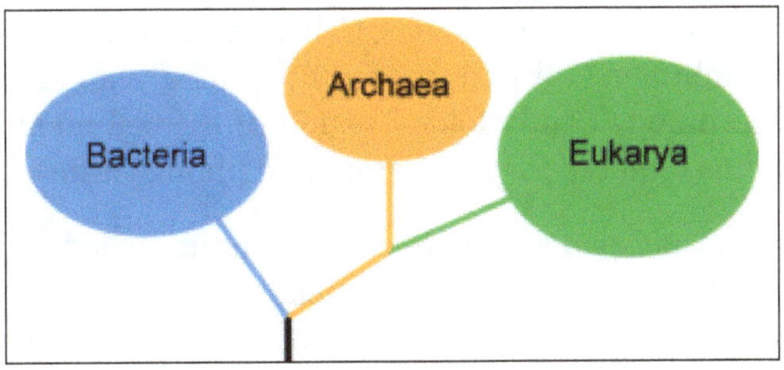

Within these domains are six kingdoms of life, these kingdoms are: Archaebacteria, Eubacteria, Protista, Fungi, Plantae and Animalia.

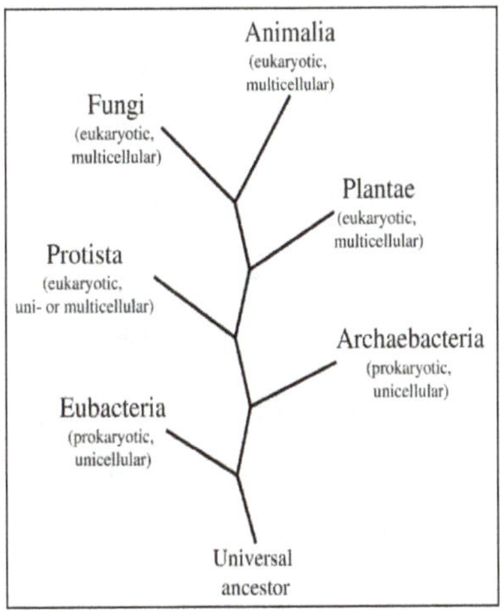

Prokaryote vs. Eukaryote

Prokaryotes (pro=before, karyon= nucleus) are single-celled organisms they lack a membrane bounded nucleus and a membrane bounded organelles, the only membrane in prokaryote cell is the plasma membrane which represent the outer boundary of the cell itself. Their genetic material is naked within the cytoplasm, the only type of organelles in prokaryote are Ribosomes. Prokaryote include two kingdoms: Archaebacteria and Eubacteria.

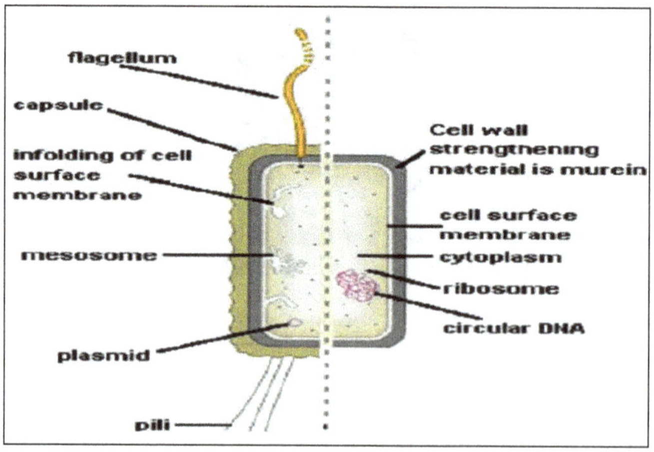

Cytoplasm of prokaryote contains all the enzymes needed for all metabolic reaction since there are no organelles. Ribosomes are the smaller (70s) type. Nuclear material is the region of cytoplasm that contains DNA. It is not surrounded by a nuclear membrane. DNA is always circular and not associated with any protein to form chromosomes. Cell membrane made of phospholipid and proteins like in eukaryotic membrane. Cell wall: made of peptidoglycan (glycoprotein= protein-carbohydrate complex). Mesosome is tightly folded region of the cell membrane containing all the membrane bounded proteins required for respiration and photosynthesis.

Eukaryote (Eu=true, karyon=nucleus) are single-celled or multi-celled organisms with a definitive nucleus and a membrane bounded organelles, organelles such as mitochondria and chloroplast. These vital organelles involved in metabolism and energy conversion within the cell. Eukaryote included four kingdoms: Protista, Fungi, Plantae and Animalia.

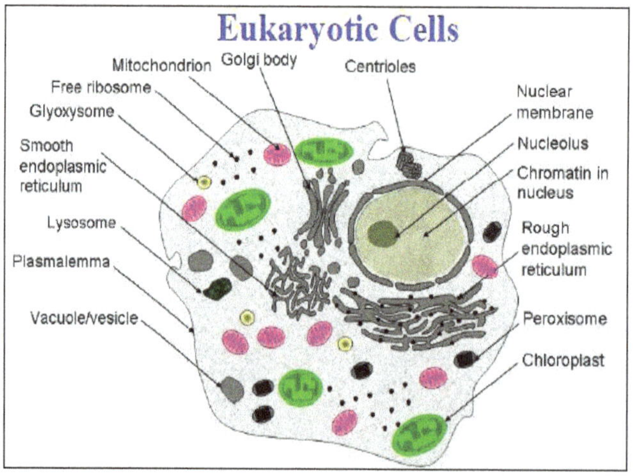

Cell membrane thin layer of phospholipid and protein, it separates the content of cell from the outside environment and control the entry and exit of material. Cytoplasm of eukaryote contains nucleus and all organelles needed by cell to function. Nucleus surrounded by double membrane nuclear envelope inside nucleus chromosomes this consists of DNA and protein.

Prokaryotic Cells	Eukaryotic cells
Small cells (< 5 μm)	Larger cells (> 10 μm)
Always unicellular	Often multicellular
No nucleus or any membrane-bound organelles, such as mitochondria	Always have nucleus and other membrane-bound organelles
DNA is circular, without proteins	DNA is linear and associated with proteins to form chromatin
Ribosomes are small (70S)	Ribosomes are large (80S)
No cytoskeleton	Always has a cytoskeleton
Motility by rigid rotating flagellum (made of flagellin)	Motility by flexible waving cilia or flagellae (made of tubulin)
Cell division is by binary fission	Cell division is by mitosis or meiosis
Reproduction is always asexual	Reproduction is asexual or sexual
Huge variety of metabolic pathways	Common metabolic pathways

All living things fall into one of three domains which include six kingdoms:

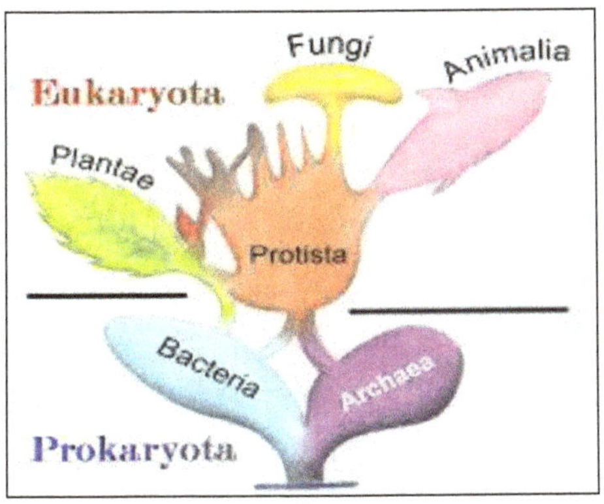

- Kingdom Archaebacteria: Prokaryotes, unicellular, that lack a peptidoglycan cell wall, autotrophic or heterotrophic, including the methanogens and extreme halophiles and thermophiles.
- Kingdom Eubacteria: Prokaryotic organisms with a peptidoglycan cell wall, autotrophic or heterotrophic, including cyanobacteria, soil bacteria, nitrogen-fixing bacteria, and pathogenic bacteria.
- Kingdom Protista: Eukaryotic, primarily unicellular and less are multicellular, autotrophic (photosynthetic) or heterotrophic organisms, such as amoebas, algae and paramecium.
- Kingdom Fungi: Eukaryotic, mostly multicellular, heterotrophic, usually nonmotile organisms, with cell walls of chitin, such as mushrooms.
- Kingdom Plantae: Eukaryotic, multicellular, nonmotile, usually terrestrial, photosynthetic organisms, such as trees, grasses, and mosses.

- Kingdom Animalia: Eukaryotic, multicellular, motile, heterotrophic organisms, such as sponges, spiders, newts and humans. As more is learned about living things, particularly from the newer evidence that DNA studies provide, scientists will continue to reevaluate the relationships among the kingdoms of life.

Bacteria

Bacteria are the oldest, structurally simplest, and the most abundant forms of life on earth. They are also the only organisms with prokaryotic cellular organization.

Bacterial Shapes and Arrangement

Bacteria are mostly simple in form and exhibit one of three basic structures:

- Coccus,
- Bacillus,
- Spiral.

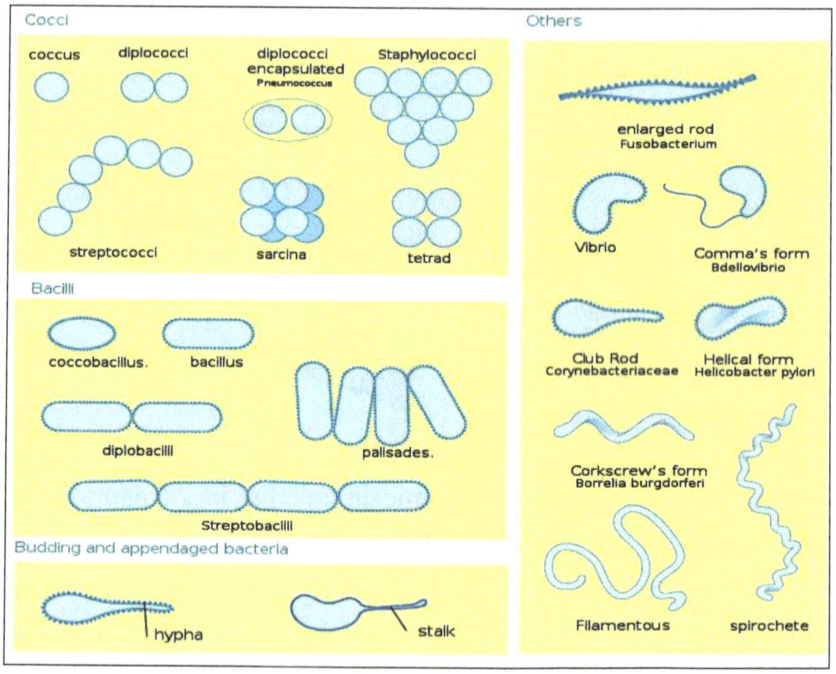

Bacterial Structure

The bacterial cell wall is an important structure because it maintains the shape of the cell and protects the cell from swelling and rupturing. The cell wall usually consists of peptidoglycan. In some bacteria, the peptidoglycan forms a thick, complex network around the outer surface of the cell. In other bacteria a thin layer of peptidoglycan is found sandwiched between two plasma membranes. The outer membrane contains large molecules of lipopolysaccharide, lipids with polysaccharide chains attached. These two major types of bacteria can be identified using a staining process called a Gram stain.

Gram-positive bacteria have the thicker peptidoglycan wall and stain a purple color. The more common gram-negative bacteria contain less peptidoglycan and do not retain the purple-colored dye. Gram-negative bacteria stain red.

In some kinds of bacteria, an additional layer consists of polysaccharide, polypeptides or both called glycocalyx when this layer highly organized and firmly attached to the wall it is called the capsule when less organized loosely attached to the wall it is called slimy layer. This gelatinous layer helps a bacterium to attach to teeth, mucus membrane, rocks in streams and other interesting surfaces; it also helps some encapsulated types to resist being engulfed by phagocytic cells of the host.

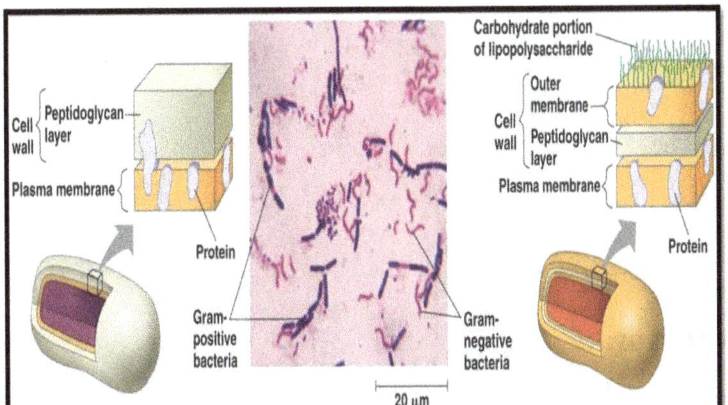

(a) Gram-positive: peptidoglycan traps crystal violet.

(b) Gram-negative: crystal violet is easily rinsed away, revealing red dye.

Many kinds of bacteria have slender, rigid, helical flagella (singular, flagellum) composed of the protein flagellin. These flagella range from 3 to 12 μm in length and are very thin only 10 to 20 nm thick. They are anchored in the cell wall and spin, pulling the bacteria through the water like a propeller. Pili (singular pilus) are other hair like structures that occur on the cells of some bacteria. They are shorter than bacterial flagella, up to several μm long and about 7.5 to 10 nm thick. Pili help the bacterial cells attach to appropriate substrates and exchange genetic information.

Bacteria lack nuclei and do not possess the complex chromosomes characteristic of eukaryotes. Instead, their genes are encoded within a single double-stranded ring of DNA that is crammed into one region of the cell known as the nucleoid region. Many bacterial cells also possess small, independently replicating circles of DNA called plasmids. Plasmids contain only a few genes, usually not essential for the cell's survival. They are best thought of as an excised portion of the bacterial chromosome. Ribosomes: Bacterial ribosomes are smaller than those of eukaryotes and differ in protein.

Bacterial Metabolism

Bacteria have evolved many mechanisms to acquire the energy and nutrients they need for growth and reproduction.

- Autotrophs: Organisms that obtain their carbon from inorganic CO_2. Autotrophs that obtain their energy from sunlight are called photoautotrophs, while those that harvest energy from inorganic chemicals are called chemoautotrophs.

- Heterotrophs: Organisms that obtain at least some of their carbon from organic molecules like glucose. Heterotrophs that obtain their energy from sunlight are called photoheterotrophs, while those that harvest energy from organic molecules are called chemoheterotrophs.

Metabolic Diversity among Organisms

Nutritional Type	Energy Source	Carbon Source	Example
Photoautotroph	Light	CO_2	Oxygenic: Cyanobacteria plants. Anoxygenic: Green, purple bacteria
Photoheterotroph	Light	Organic compounds	Green, purple nonsulfur bacteria
Chemoautotroph	Chemical	CO_2	Iron-oxidizing bacteria
Chemoheterotroph	Chemical	Organic compounds	Fermentative bacteria Animals, protozoa, fungi, bacteria.

Bacterial Reproduction

Most bacteria reproduce by binary fission or by budding (external bud from the parent cell). During binary fission, the single DNA molecule replicates and both copies attach to the cell membrane. The cell membrane begins to grow between the two DNA molecules. Once the bacterium just about doubles its original size, the cell membrane begins to pinch inward. The cell wall then forms between the two DNA molecules dividing the original cell into two identical daughter cells.

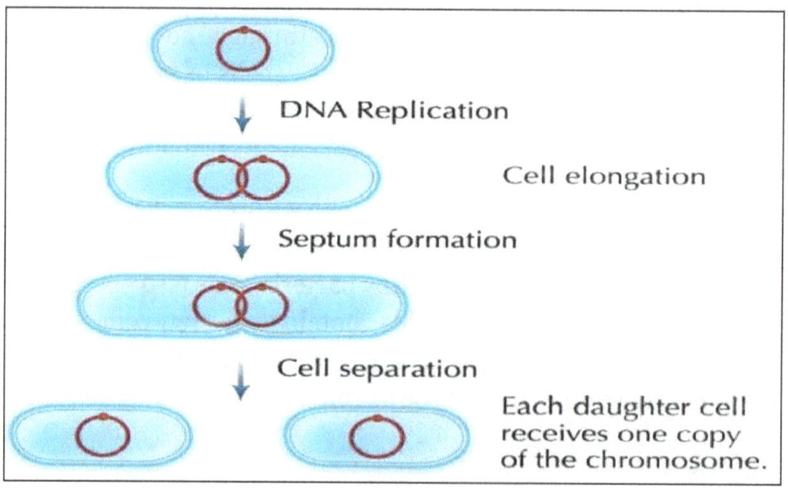

Binary fission in bacteria.

Binary fission is an effective way for bacteria to reproduce, however it does produce problems. Since the cells produced through this type of reproduction are identical, they are all susceptible to the same types of antibiotics. In order to incorporate some genetic variation, bacteria use a process called recombination. Bacterial recombination can be accomplished through conjugation, transformation, or transduction.

Conjugation

Some bacteria are capable of transferring pieces of their genes to other bacteria that they come in contact with. During conjugation, one bacterium connects itself to another through a protein tube structure called a pilus. Genes are transferred from one bacterium to the other through this tube.

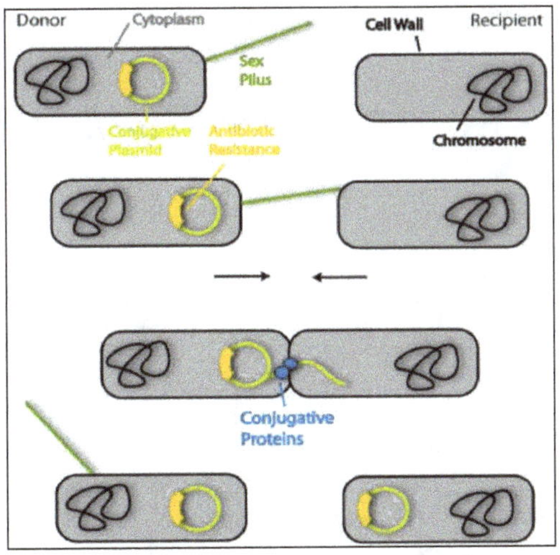

Transformation

Some bacteria are capable of taking up DNA from their environment. These DNA remnants most commonly come from dead bacterial cells. During transformation, the bacterium binds the DNA and transports it across the bacterial cell membrane. The new DNA is then incorporated into the bacterial cell's DNA.

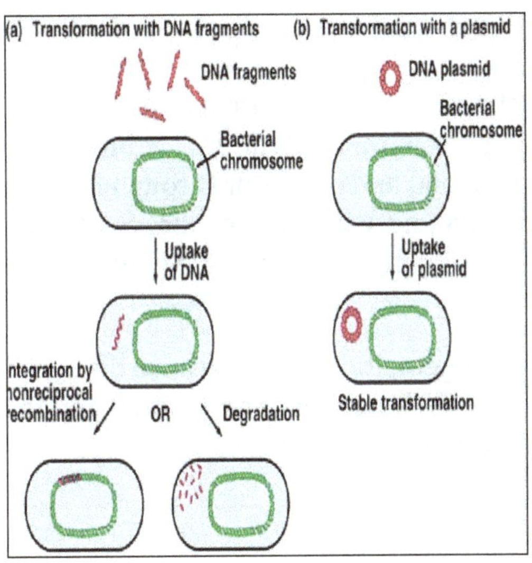

Transduction

Transduction is a type of recombination that involves the exchanging of bacterial DNA through bacteriophages. Bacteriophages are viruses that infect bacteria. There are two types of transduction: generalized and specialized transduction.

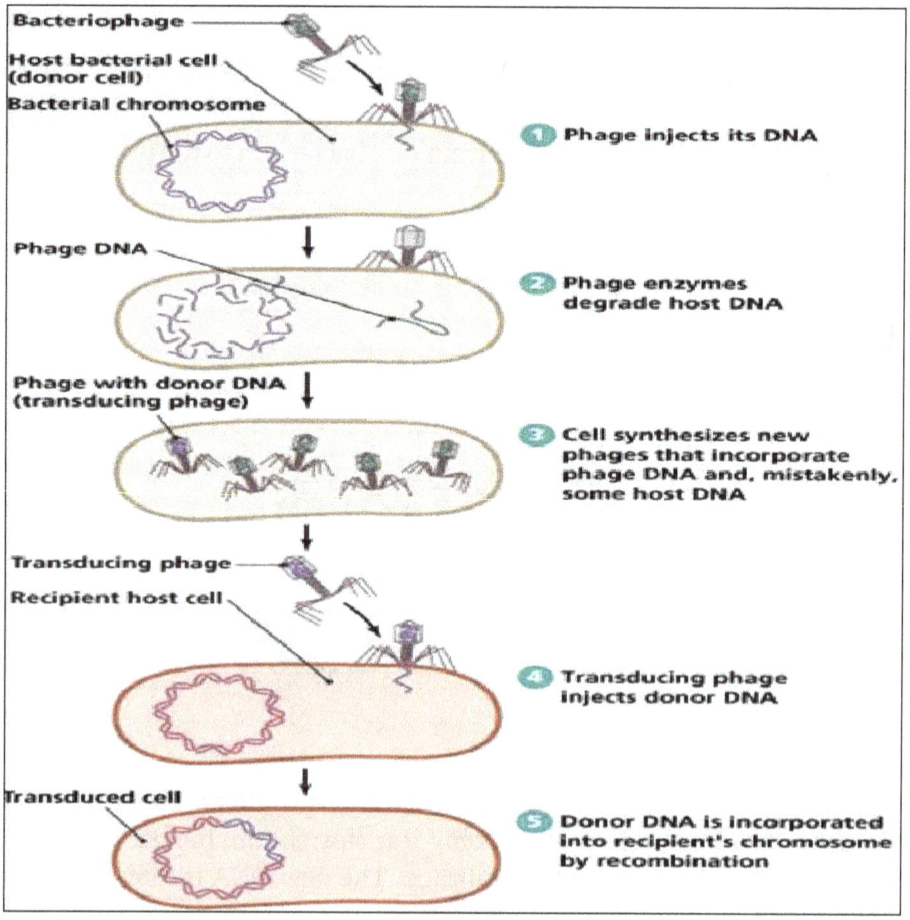

Bacteria and Disease

Although the great majority of bacteria play a positive role in our life, still there are many bacteria are pathogenic (disease causing) to human, plant and animal. In humans bacteria cause many diseases acquired with consumed food and drinks or through inhalation such: Cholera, Dysentery, Tuberculosis, Bacterial Pneumonia and whooping cough. Some bacterial species responsible for some sexually transmitted diseases such as Gonorrhea, Syphilis which are acquired through sexual intercourse.

Bacteria vary greatly in their pathogenicity (ability to cause disease) virulence (degree of pathogenicity), invasiveness (ability to invade host tissues and fluids) and toxicity (ability to create clinical response due to exotoxins and endotoxins).

Viruses

A virus (noncellular infectious particle) consists of nucleic acid (either DNA or RNA), a protein coat or sheath or capsid (one or more types of protein subunits organized in to a rod like or much side shape) and sometimes an outer envelope (mostly of membrane remnants from a previously infected cell). Virus cannot reproduce itself. It can be reproducing only after its genetic material and few enzymes enter a host cell and subvert the cells biosynthetic machinery.

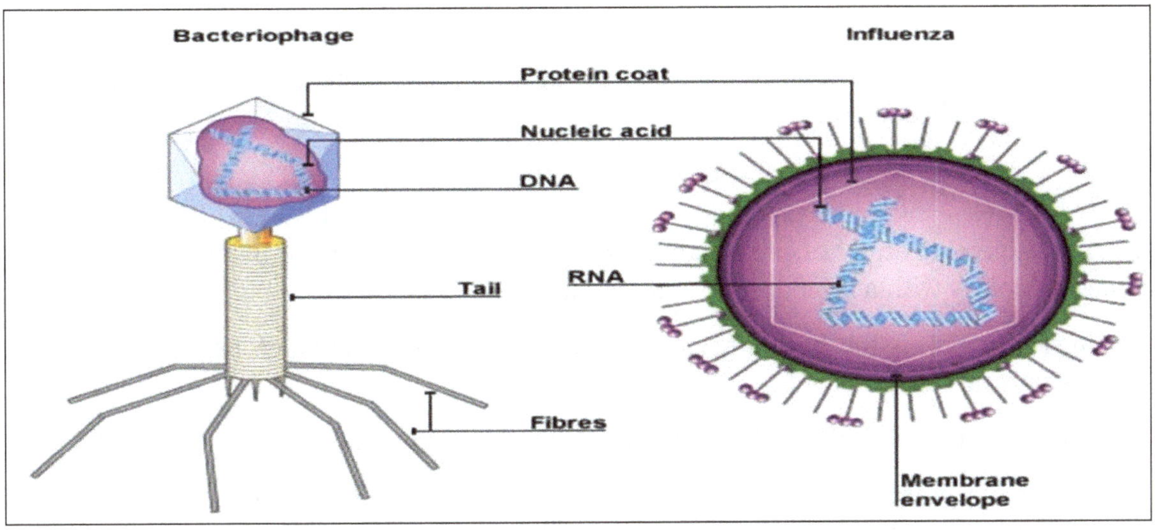

The size of virus is usually less than 0.3 μm. However, some viruses form large particle which can be seen under high power light microscope. Viruses cause many diseases in all living organisms such as:

- Smallpox,
- Chicken pox,
- Measles,
- German measles (Rubella),
- Viral encephalitis,
- Mumps,
- Influenza,
- Infectious Hepatitis,
- Yellow fever,
- Rabies,
- AIDS.

In addition viruses have been implicated in some cancers and leukemia.

Nearly all viral multiplication cycles include five steps:

- Attachment to suitable host cell,
- Penetration of it,
- DNA or RNA replication,
- Protein synthesis,

- Assembly of new viral particles and release.

Bacteriophages are a group of viruses that infect bacterial cells. Some bacteriophages consist only of DNA and a protein coat.

Branches of Biology

Biology, the scientific study of life, includes several relevant branches. Below is a list of major branches of biology with a brief description for each:

- Agriculture: It is the study of science and practice of producing crops and livestock from the natural resources of the earth.
- Anatomy: It is the study of the animal form, particularly human body.
- Astrobiology: It is the branch of biology concerned with the effects of outer space on living organisms and the search for extraterrestrial life.
- Biochemistry: It is the study of the structure and function of cellular components, such as proteins, carbohydrates, lipids, nucleic acids, and other biomolecules, and of their functions and transformations during life processes.
- Bioclimatology: It is the study of science concerned with the influence of climates on organisms, for instance the effects of climate on the development and distribution of plants, animals, and humans.
- Bioengineering or Biological Engineering: It is a broad-based engineering discipline that deals with bio-molecular and molecular processes, product design, sustainability and analysis of biological systems.
- Biogeography: It is the study of science that attempts to describe the changing distributions and geographic patterns of living and fossil species of plants and animals.
- Bioinformatics: It is an information technology applied to the life sciences, especially the technology used for the collection, storage, and retrieval of genomic data.
- Biomathematics: It is the mathematical biology or biomathematics is an interdisciplinary field of academic study which aims at modeling natural, biological processes using mathematical techniques and tools. It has both practical and theoretical applications in biological research.
- Biophysics of Biological Physics: It is an interdisciplinary science that applies the theories and methods of physical sciences to questions of biology.
- Biotechnology: It is the study of applied science that is concerned with biological systems, living organisms, or derivatives thereof, to make or modify products or processes for specific use.
- Botany: It is the scientific study of plants.
- Cell Biology: It is the study of cells at the microscopic or at the molecular level. It includes studying the cells' physiological properties, structures, organelles, interactions with their environment, life cycle, division and apoptosis.

- Chronobiology: It is a science that studies time-related phenomena in living organisms.
- Conservation Biology: It is concerned with the studies and schemes of habitat preservation and species protection for the purpose of alleviating extinction crisis and conserving bio-diversity.
- Cryobiology: It is the study of the effects of low temperatures on living organisms.
- Developmental Biology: It is the study of the processes by which an organism develops from a zygote to its full structure.
- Ecology: It is the scientific study of the relationships between plants, animals, and their environment.
- Ethnobiology: It is a study of the past and present human interactions with the environment, for instance the use of diverse flora and fauna by indigenous societies.
- Evolutionary Biology: It is a subfield concerned with the origin and descent of species, as well as their change over time, i.e. their evolution.
- Freshwater Biology: It is a science concerned with the life and ecosystems of freshwater habitats.
- Genetics: It is a science that deals with heredity, especially the mechanisms of hereditary transmission and the variation of inherited characteristics among similar or related organisms.
- Geobiology: It is a science that combines geology and biology to study the interactions of organisms with their environment.
- Immunobiology: It is a study of the structure and function of the immune system, innate and acquired immunity, the bodily distinction of self from oneself, and laboratory techniques involving the interaction of antigens with specific antibodies.
- Marine Biology: It is the study of ocean plants and animals and their ecological relationships.
- Medicine: It is the study of science which relates to the prevention, cure, or alleviation of disease.
- Microbiology: It is the branch of biology that deals with microorganisms and their effects on other living organisms.
- Molecular Biology: It is the branch of biology that deals with the formation, structure, and function of macromolecules essential to life, such as nucleic acids and proteins, and especially with their role in cell replication and the transmission of genetic information.
- Mycology: It is the study of fungi.
- Neurobiology: It is the branch of biology that deals with the anatomy and physiology and pathology of the nervous system.
- Paleobiology: It is the study of the forms of life existing in prehistoric or geologic times, as represented by the fossils of plants, animals, and other organisms.

- Parasitology: It is the study of parasites and parasitism.
- Pathology: It is the study of the nature of disease and its causes, processes, development, and consequences.
- Pharmacology: It is the study of preparation and use of drugs and synthetic medicines.
- Physiology: It is the biological study of the functions of living organisms and their parts.
- Protistology: It is the study of protists.
- Psychobiology: It is the study of mental functioning and behavior in relation to other biological processes.
- Toxicology: It is the study of how natural or man-made poisons cause undesirable effects in living organisms.
- Virology: It is study of viruses.
- Zoology: It is the branch of biology that deals with animals and animal life, including the study of the structure, physiology, development, and classification of animals.
- Ethology: It is the study of animal behavior.
- Entomology: It is the scientific study of insects.
- Ichthyology: It is the study of fishes.
- Herpetology: It is the study of reptiles and amphibians.
- Ornithology: It is the study of birds.
- Mammalogy: It is the study of mammals
- Primatology: It is the study of science that deals with primates.

Chapter 2

Genes and Cells

Cells are the fundamental units of life. The ability of cells to divide led to the creation of advanced multicellular life forms. Contained within these cells are genes, through which traits are passed down from one generation to the next. This chapter seeks to explain the topic in a simple and detailed manner for the reader's benefit.

Genes

Genes are functional units of heredity as they are made of DNA. The chromosome is made of DNA containing many genes. Every gene comprises of the particular set of instructions for a particular function or protein-coding. Speaking in usual terms, genes are responsible for heredity. There are about 30000 genes in each cell of the human body. DNA present in the gene comprises only 2 percent of the genome. Many studies have been made on the same that found the location of nearly 13000 genes on each of the chromosomes.

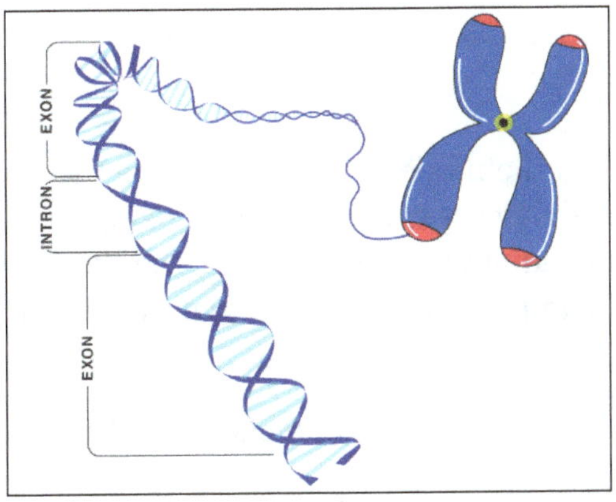

William Bateson introduced the term genetics in the year 1905. Later, Wilhelm Johannes was the first one who coined the term GENE in 1909. He was a Danish botanist. He named it as Gene to symbolize hereditary.

Characteristics Determined by Genes

The human cell contains 23 pairs of chromosomes. The trait is one of the characteristics determined by one or more genes. Abnormal genes and genes that are formed due to new mutations also result in certain traits. Genes vary in size depending on the code or the protein they produce. All cells in the human body contain the same DNA. The difference between the cells occurs due to the different type of genes that are turned on and therefore produce a variety of proteins.

Reasons for Hereditary

Genes come in pairs in the same way as the chromosomes. Each parent of a human being carries two copies of their genes and each parent passes one copy of genes to their child. This is the reason why the child has many characteristics of both the parents like hair colour, same eyes etc.

Functions of Genes

- Genes control the functions of DNA and RNA.
- Proteins are the most important materials in the human body which not only help by being the building blocks for muscles, connecting tissue and skin but also take care of the production of the enzyme.
- These enzymes play an important role in conducting various chemical processes and reactions within the body. Therefore, protein synthesis is responsible for all activities carried on by the body and are mainly controlled by the genes.
- Genes consist of a particular set of instructions or specific functions. For example, the globin gene was instructed to produce hemoglobin. Hemoglobin is a protein that helps to carry oxygen in the blood.

Cells

Cells are the basic unit of life. In the modern world, they are the smallest known world that performs all of life's functions. All living organisms are either single cells, or are multicellular organisms composed of many cells working together. Cells are the smallest known unit that can accomplish all of these functions. Defining characteristics that allow a cell to perform these functions include:

- A cell membrane that keeps the chemical reactions of life together.
- At least one chromosome composed of genetic material that contains the cell's "blueprints" and "software".
- Cytoplasm: The fluid inside the cell, in which the chemical processes of life occur.

Function of Cells

Scientists define seven functions that must be fulfilled by a living organism. These are:

- A living thing must respond to changes in its environment.
- A living thing must grow and develop across its lifespan.
- A living thing must be able to reproduce, or make copies itself.

- A living thing must have metabolism.
- A living thing must maintain homeostasis, or keep its internal environment the same regardless of outside changes.
- A living thing must be made of cells.
- A living thing must pass on traits to its offspring.

It is the biology of cells which enables living things to perform all of these functions.

How Cells Work

In order to accomplish them, they must have:

- A cell membrane that separates the inside of the cell from the outside. By concentrating the chemical reactions of life inside a small area within a membrane, cells allow the reactions of life to proceed much faster than they otherwise would.
- Genetic material which is capable of passing on traits to the cell's offspring. In order to reproduce, organisms must ensure that their offspring have all the information that they need to be able to carry out all the functions of life. All modern cells accomplish this using DNA, whose base-pairing properties allow cells to make accurate copies of a cell's "blueprints" and "operating system." Some scientists think that the first cells might have used RNA instead.
- Proteins that perform a wide variety of structural, metabolic, and reproductive functions. There are countless different functions that cells must perform to obtain energy and reproduce.

Depending on the cell, examples of these functions can include photosynthesis, breaking down sugar, locomotion, copying its own DNA, allowing certain substances to pass through the cell membrane while keeping others out, etc.

Proteins are made of amino acids, which are like the "Legos" of biochemistry. Amino acids come in different sizes, different shapes, and with different properties such as polarity, ionic charge, and hydrophobicity.

By putting amino acids together based on the instructions in their genetic material, cells can create biochemical machinery to perform almost any function. Some scientists think that the first cells might have used RNA to accomplish some vital functions, and then moved to much more versatile amino acids to do the job as the result of a mutation.

Cell Types

Because of the millions of diverse species of life on Earth, which grow and change gradually over time, there are countless differences between the countless extant types of cells.

Prokaryotes

Prokaryotes are the simpler and older of the two major types of cells. Prokaryotes are single-celled organisms. Bacteria and archaebacteria are examples of prokaryotic cells.

Prokaryotic cells have a cell membrane, and one or more layers of additional protection from the outside environment. Many prokaryotes have a cell membrane made of phospholipids, enclosed by a cell wall made of a rigid sugar. The cell wall may be enclosed by another thick "capsule" made of sugars. Many prokaryotic cells also have cilia, tails, or other ways in which the cell can control its movement.

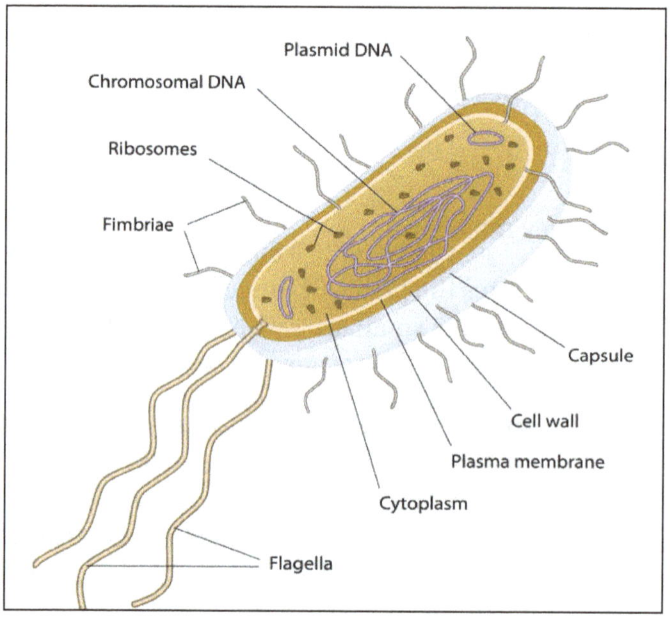

Prokaryote cell.

These characteristics, as well as the cell wall and capsule, reflect the fact that prokaryotic cells are going it alone in the environment. They are not part of a multicellular organism, which might have whole layers of cells devoted to protecting other cells from the environment, or to creating motion.

Prokaryotic cells have a single chromosome which contains all of the cell's essential hereditary material and operating instructions. This single chromosome is usually round. There is no nucleus, or any other internal membranes or organelles. The chromosome just floats in the cell's cytoplasm. Additional genetic traits and information might be contained in other gene units within the cytoplasm, called "plasmids," but these are usually genes that are passed back and forth by prokaryotes though the process of "horizontal gene transfer," which is when one cell gives genetic material to another. Plasmids contain non-essential DNA that the cell can live without, and which is not necessarily passed on to offspring.

When a prokaryotic cell is ready to reproduce, it makes a copy of its single chromosome. Then the cell splits in half, apportioning one copy of its chromosome and a random assortment of plasmids to each daughter cell. There are two major types of prokaryotes known to scientists to date: archaebacteria, which are a very old lineage of life with some biochemical differences from bacteria and eukaryotes, and bacteria, sometimes called "eubacteria," or "true bacteria" to differentiate them from archaebacteria.

Bacteria are thought to be more "modern" descendants of archaebacteria. Both families have "bacteria" in the name because the differences between them were not understood prior to the invention of modern biochemical and genetic analysis techniques. When scientists began to examine the biochemistry and genetics of prokaryotes in detail, they discovered these two very different groups, who probably have different relationships to eukaryotes and different evolutionary histories.

Some scientists think that eukaryotes like humans are more closely related to bacteria, since eukaryotes have similar cell membrane chemistry to bacteria. Others think that archaebacteria are more closely related to us eukaryotes, since they use similar proteins to reproduce their chromosomes.

Still others think that we might be descended from both – that eukaryotic cells might have come into existence when archaebacteria started living inside of a bacterial cell, or vice versa. This would explain how we have important genetic and chemical attributes of both, and why we have multiple internal compartments such as the nucleus, chloroplasts, and mitochondria.

Eukaryotes

Eukaryotic cells are thought to be the most modern major cell type. All multicellular organisms, including you, your cat, and your houseplants, are eukaryotes. Eukaryotic cells seem to have "learned" to work together to create multicellular organisms, while prokaryotes seem unable to do this. Eukaryotic cells usually have more than one chromosome, which contains large amounts of genetic information. Within the body of a multicellular organism, different genes within these chromosomes may be switched "on" and "off," allowing for cells that have different traits and perform different functions within the same organism.

Eukaryotic cells also have one or more internal membranes, which has led scientists to the conclusion that eukaryotic cells likely evolved when one or more types of prokaryote began living in symbiotic relationships inside of other cells. Organelles with interior membranes found in eukaryotic cells typically include:

- For animal cells: Mitochondria, which liberate the energy from sugar and turn it into ATP in an extremely efficient way. Mitochondria even have their own DNA, separate from the cells' nuclear DNA, which gives further support for the theory that they used to be independent bacteria.
- For plant cells: Chloroplasts, which perform photosynthesis, making ATP and sugar from sunlight and air. Chloroplasts also have their own DNA, suggesting that they may have originated as photosynthetic bacteria.
- Nucleus: In eukaryotic cells, the nucleus contains the essential DNA blueprints and operating instructions for the cell. The nuclear envelope is thought to provide an extra layer of protection for the DNA against toxins or invaders which might damage it. It is unknown whether the nucleus might also have been an endosymbiotic prokaryote at one time, or whether its membrane simply evolved as an extra layer of protection for the cell's DNA.
- Endoplasmic reticulum: This complex internal membrane is a major site of protein creation for cells. The evolutionary origin of the endoplasmic reticulum is not known.
- Golgi apparatus: This internal membrane complex can be thought of like the endoplasmic reticulum's "post office." It receives proteins from the ER, packages and "labels" them by attaching sugars as needed, and then ships them off to their final destinations.

- Others: Many eukaryotic cells can create temporary internal membrane "sacs," called "vacuoles," to store waste, or to package important materials. Some cells, for example have special vacuoles called "lysosomes" which are full of corrosive substances and digestive enzymes. Cells simply dump their "trash" into lysosomes, where the harsh environment breaks them down into simpler components that can be re-used.

Examples of Cells

Archaebacteria

Archaebacteria are a very old form of prokaryotic cells. Biologists actually put them in their own "domain" of life, separate from other bacteria. Key ways in which archaebacteria differ from other bacteria includes:

- Their cell membranes, which are made of a type of lipid not found in either bacteria or eukaryotic cell membranes.
- Their DNA replication enzymes, which are more similar to those of eukaryotes than those of bacteria, suggesting that bacteria and archae are only distantly related, and archaebacteria may actually be more closely related to us than to modern bacteria.
- Some archaebacteria have the ability to produce methane, which is a metabolic process not found in any bacteria or any eukaryotes.

Archaebacteria's unique chemical attributes allow them to live in extreme environments, such as superheated water, extremely salty water, and some environments which are toxic to all other life forms. Scientists became very excited in recent years at the discovery of Lokiarchaeota – a type of archaebacteria which shares many genes with eukaryotes that had never before been found in prokaryotic cells.

Bacteria

You are most likely familiar with the type of bacteria that can make you sick. Indeed, common pathogens like Streptococcus and Staphylococcus are prokaryotic bacterial cells. But there are also many types of helpful bacteria – including those that break down dead waste to turn useless materials into fertile soil, and bacteria. Bacterial cells can commonly be found living in symbiotic relationships with multicellular organisms like us, in the soil, and anywhere else that's not too extreme for them to live.

Plant Cells

Plant cells are eukaryotic cells that are part of multicellular, photosynthetic organisms. Plants cells have chloroplast organelles, which contain pigments that absorb photons of light and harvest the energy of those photons.

Chloroplasts have the remarkable ability to turn light energy into cellular fuel, and use this energy to take carbon dioxide from the air and turn it into sugars that can be used by living things as fuel or building material. In addition to having chloroplasts, plant cells also typically have a cell wall made of a rigid sugars, to enable plant tissues to maintain their upright structures such as leaves, stems, and tree trunks. Plant cells also have the usual eukaryotic organelles including a nucleus, endoplasmic reticulum, and Golgi apparatus.

Animal Cells

For this exercise, let's look at a type of animal cell that is of great importance to you: your own liver cell:

- Like all animal cells, it has mitochondria which perform cellular respiration, turning oxygen and sugar into large amounts of ATP to power cellular functions.
- It also has the same organelles as most animal cells: a nucleus, endoplasmic reticulum, Golgi apparatus, etc.
- But as part of a multicellular organism, your liver cell also expresses unique genes, which give it unique traits and abilities.
- Liver cells in particular contain enzymes that break down many toxins, which is what allows the liver to purify your blood and break down dangerous bodily waste.
- The liver cell is an excellent example of how multicellular organisms can be more efficient by having different cell types work together.

Cell Membrane

The structure and function of cells are critically dependent on membranes, which not only separate the interior of the cell from its environment but also define the internal compartments of eukaryotic cells, including the nucleus and cytoplasmic organelles. The formation of biological membranes is based on the properties of lipids, and all cell membranes share a common structural organization: bilayers of phospholipids with associated proteins. These membrane proteins are responsible for many specialized functions; some act as receptors that allow the cell to respond to external signals, some are responsible for the selective transport of molecules across the membrane, and others participate in electron transport and oxidative phosphorylation. In addition, membrane proteins control the interactions between cells of multicellular organisms. The common structural organization of membranes thus underlies a variety of biological processes and specialized membrane functions.

Membrane Lipids

The fundamental building blocks of all cell membranes are phospholipids, which are amphipathic molecules, consisting of two hydrophobic fatty acid chains linked to a phosphate-containing hydrophilic head group. Because their fatty acid tails are poorly soluble in water, phospholipids spontaneously form bilayers in aqueous solutions, with the hydrophobic tails buried in the interior of the membrane and the polar head groups exposed on both sides. Such phospholipid bilayers form a stable barrier between two aqueous compartments and represent the basic structure of all biological membranes.

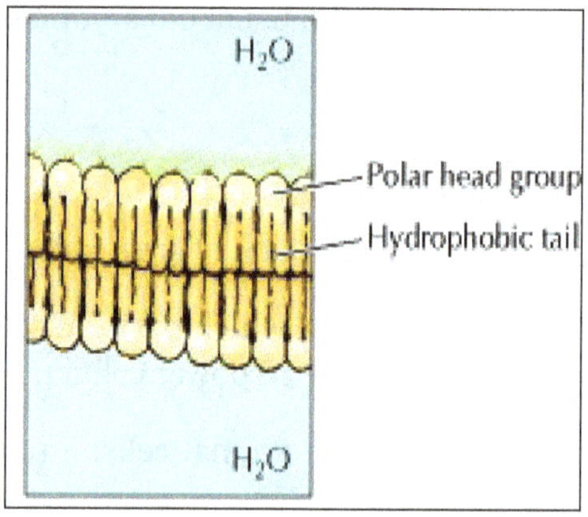

A phospholipid bilayer: Phospholipids spontaneously form stable bilayers, with their polar head groups exposed to water and their hydrophobic tails buried in the interior of the membrane.

Lipids constitute approximately 50% of the mass of most cell membranes, although this proportion varies depending on the type of membrane. Plasma membranes, for example, are approximately 50% lipid and 50% protein. The inner membrane of mitochondria, on the other hand, contains an unusually high fraction (about 75%) of protein, reflecting the abundance of protein complexes involved in electron transport and oxidative phosphorylation. The lipid composition of different cell membranes also varies. The plasma membrane of E. coli consists predominantly of phosphatidylethanolamine, which constitutes 80% of total lipid. Mammalian plasma membranes are more complex, containing four major phospholipids—phosphatidylcholine, phosphatidylserine, phosphatidylethanolamine, and sphingomyelin—which together constitute 50 to 60% of total membrane lipid. In addition to the phospholipids, the plasma membranes of animal cells contain glycolipids and cholesterol, which generally correspond to about 40% of the total lipid molecules.

	Plasma membrane			
Lipid	**E. coli**	**Erthrocyte**	**Rough endoplasmic reticulum**	**Outer mitochondrial membranes**
Phosphatidylcholine	0	17	55	50
Phosphatidylserine	0	6	3	2
Phosphatidylethanolamine	80	16	16	23
Sphingomyelin	0	17	3	5
Glycolipids	0	2	0	0
Cholesterol	0	45	6	<5

An important property of lipid bilayers is that they behave as two-dimensional fluids in which individual molecules (both lipids and proteins) are free to rotate and move in lateral directions as shown in figure below. Such fluidity is a critical property of membranes and is determined by both temperature and lipid composition. For example, the interactions between shorter fatty acid chains are weaker than those between longer chains, so membranes containing shorter fatty acid chains are less rigid and remain fluid at lower temperatures. Lipids containing unsaturated fatty acids similarly increase membrane fluidity because the presence of double bonds introduces kinks in the fatty acid chains, making them more difficult to pack together.

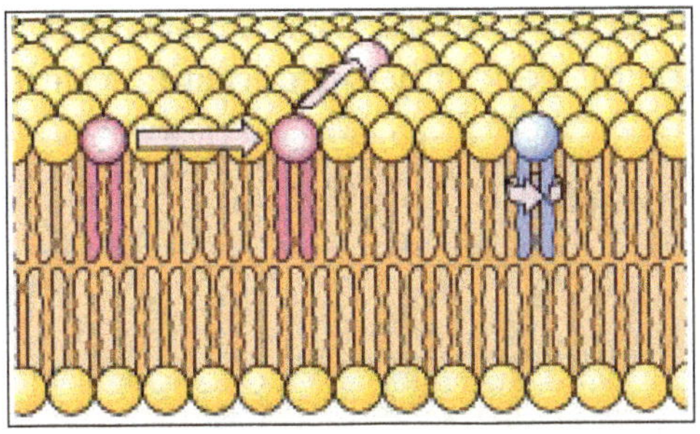

Mobility of phospholipids in a membrane. Individual phospholipids can rotate and move laterally within a bilayer.

Because of its hydrocarbon ring structure, cholesterol plays a distinct role in determining membrane fluidity. Cholesterol molecules insert into the bilayer with their polar hydroxyl groups close to the hydrophilic head groups of the phospholipids in figure below. The rigid hydrocarbon rings of cholesterol therefore interact with the regions of the fatty acid chains that are adjacent to the phospholipid head groups. This interaction decreases the mobility of the outer portions of the fatty acid chains, making this part of the membrane more rigid. On the other hand, insertion of cholesterol interferes with interactions between fatty acid chains, thereby maintaining membrane fluidity at lower temperatures.

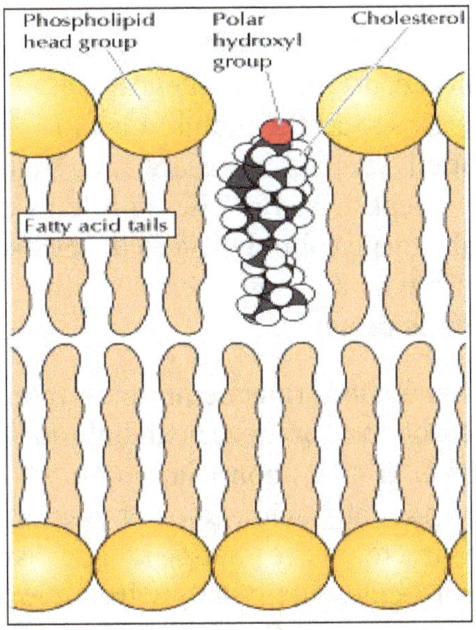

Insertion of cholesterol in a membrane. Cholesterol inserts into the membrane with its polar hydroxyl group close to the polar head groups of the phospholipids.

Membrane Proteins

Proteins are the other major constituent of cell membranes, constituting 25 to 75% of the mass of the various membranes of the cell. The current model of membrane structure, proposed by Jonathan Singer and Garth Nicolson in 1972, views membranes as a fluid mosaic in which proteins are inserted into a lipid bilayer in figure below. While phospholipids provide the basic structural organization of membranes, membrane proteins carry out the specific functions of the different membranes of the cell. These proteins are divided into two general classes, based on the nature of their association with the membrane. Integral membrane proteins are embedded directly within the lipid bilayer. Peripheral membrane proteins are not inserted into the lipid bilayer but are associated with the membrane indirectly, generally by interactions with integral membrane proteins.

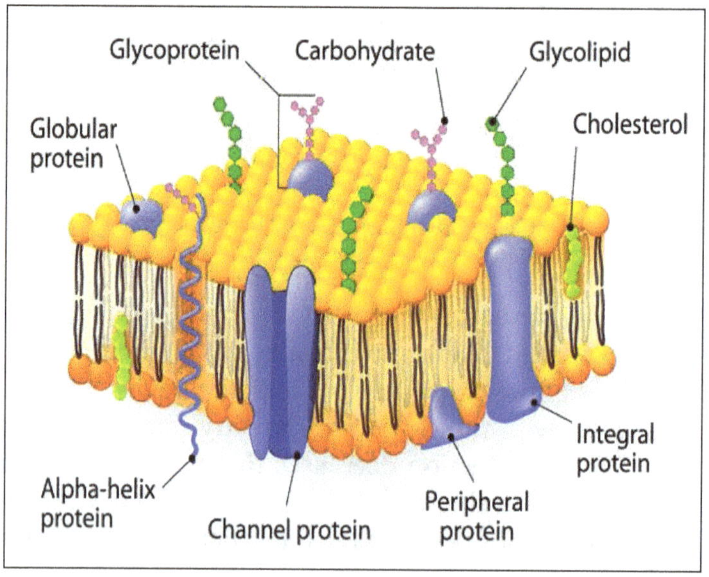

Fluid mosaic model of membrane structure.

Biological membranes consist of proteins inserted into a lipid bilayer. Integral membrane proteins are embedded in the membrane, usually via α-helical regions of 20 to 25 hydrophobic amino acids. Some transmembrane proteins span the membrane only once; others have multiple membrane-spanning regions. In addition, some proteins are anchored in the membrane by lipids that are covalently attached to the polypeptide chain. These proteins can be anchored to the extracellular face of the plasma membrane by glycolipids and to the cytosolic face by fatty acids or prenyl groups. Peripheral membrane proteins are not inserted in the membrane but are attached via interactions with integral membrane proteins.

Many integral membrane proteins (called transmembrane proteins) span the lipid bilayer, with portions exposed on both sides of the membrane. The membrane-spanning portions of these proteins are usually α-helical regions of 20 to 25 nonpolar amino acids. The hydrophobic side chains of these amino acids interact with the fatty acid chains of membrane lipids, and the formation of an α helix neutralizes the polar character of the peptide bonds. Like the phospholipids, transmembrane proteins are amphipathic molecules, with their hydrophilic portions exposed to the aqueous environment on both sides of the membrane. Some transmembrane proteins span the membrane only once; others have multiple membrane-spanning regions. Most transmembrane proteins of eukaryotic plasma membranes have been modified by the addition of carbohydrates, which are exposed on the surface of the cell and may participate in cell-cell interactions.

Proteins can also be anchored in membranes by lipids that are covalently attached to the polypeptide chain. Distinct lipid modifications anchor proteins to the cytosolic and extracellular faces of the plasma membrane. Proteins can be anchored to the cytosolic face of the membrane either by the addition of a 14-carbon fatty acid (myristic acid) to their amino terminus or by the addition of either a 16-carbon fatty acid (palmitic acid) or 15- or 20-carbon prenyl groups to the side chains of cysteine residues. Alternatively, proteins are anchored to the extracellular face of the plasma membrane by the addition of glycolipids to their carboxy terminus.

Transport across Cell Membranes

The selective permeability of biological membranes to small molecules allows the cell to control and maintain its internal composition. Only small uncharged molecules can diffuse freely through phospholipid bilayers in figure below. Small nonpolar molecules, such as O_2 and CO_2, are soluble in the lipid bilayer and therefore can readily cross cell membranes. Small uncharged polar molecules, such as H_2O, also can diffuse through membranes, but larger uncharged polar molecules, such as glucose, cannot. Charged molecules, such as ions, are unable to diffuse through a phospholipid bilayer regardless of size; even H^+ ions cannot cross a lipid bilayer by free diffusion.

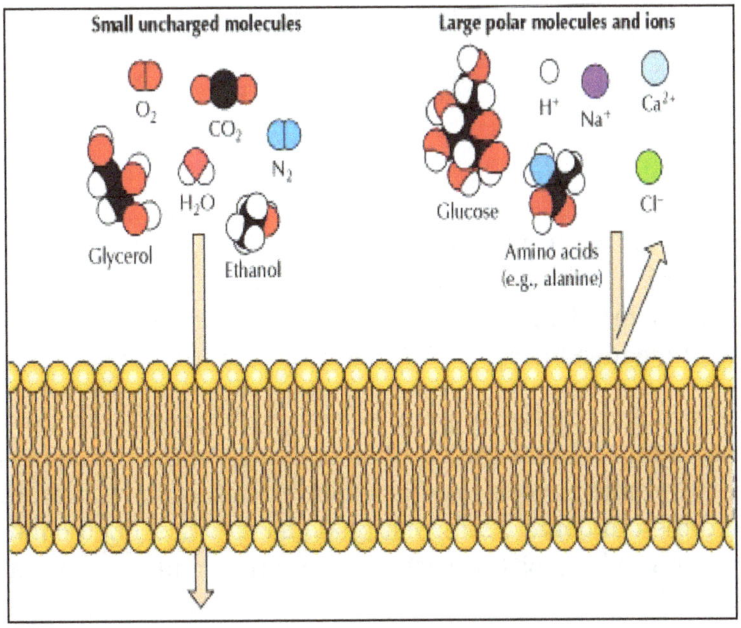

Permeability of phospholipid bilayers. Small uncharged molecules can diffuse freely through a phospholipid bilayer. However, the bilayer is impermeable to larger polar molecules (such as glucose and amino acids) and to ions.

Although ions and most polar molecules cannot diffuse across a lipid bilayer, many such molecules (such as glucose) are able to cross cell membranes. These molecules pass across membranes via the action of specific transmembrane proteins, which act as transporters. Such transport proteins determine the selective permeability of cell membranes and thus play a critical role in membrane function. They contain multiple membrane-spanning regions that form a passage through the lipid bilayer, allowing polar or charged molecules to cross the membrane through a protein pore without interacting with the hydrophobic fatty acid chains of the membrane phospholipids.

There are two general classes of membrane transport proteins as shown in figure below. Channel proteins form open pores through the membrane, allowing the free passage of any molecule of the appropriate size. Ion channels, for example, allow the passage of inorganic ions such as Na^+, K^+, Ca^{2+}, and Cl^- across the plasma membrane. Once open, channel proteins form small pores through which ions of the appropriate size and charge can cross the membrane by free diffusion. The pores formed by these channel proteins are not permanently open; rather, they can be selectively opened and closed in response to extracellular signals, allowing the cell to control the movement of ions across the membrane. Such regulated ion channels have been particularly well studied in nerve and muscle cells, where they mediate the transmission of electrochemical signals.

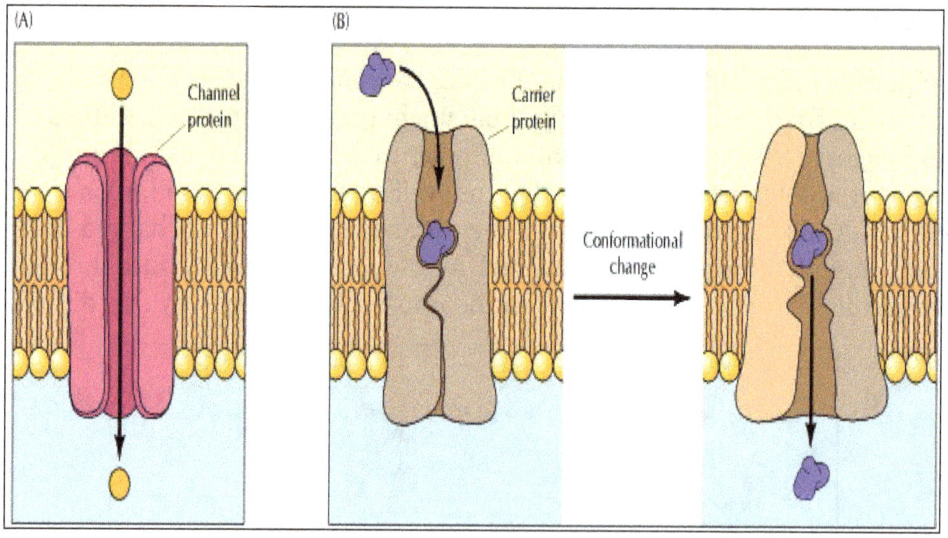

Channel and carrier proteins: (A) Channel proteins form open pores through which molecules of the appropriate size (e.g., ions) can cross the membrane. (B) Carrier proteins selectively bind the small molecule to be transported and then undergo a conformational change to release the molecule on the other side of the membrane.

In contrast to channel proteins, carrier proteins selectively bind and transport specific small molecules, such as glucose. Rather than forming open channels, carrier proteins act like enzymes to facilitate the passage of specific molecules across membranes. In particular, carrier proteins bind specific molecules and then undergo conformational changes that open channels through which the molecule to be transported can pass across the membrane and be released on the other side.

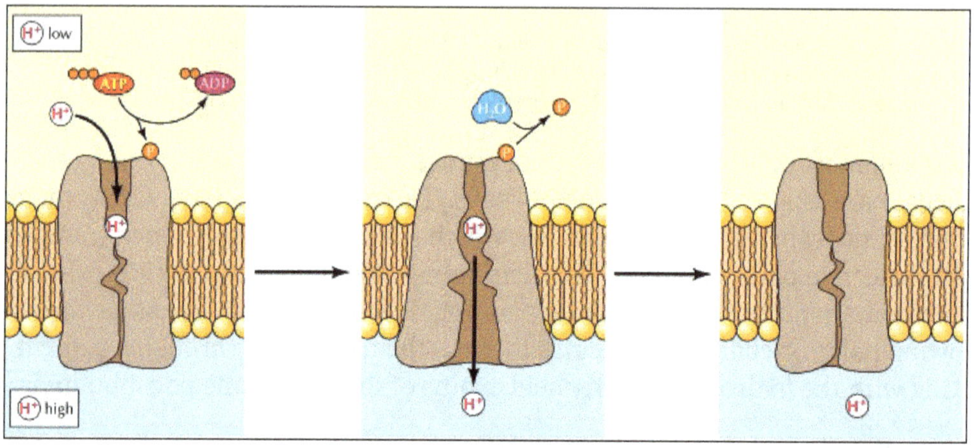

Model of active transport.

Molecules transported by either channel or carrier proteins cross membranes in the energetically favorable direction, as determined by concentration and electrochemical gradients—a process known as passive transport. However, carrier proteins also provide a mechanism through which the energy changes associated with transporting molecules across a membrane can be coupled to the use or production of other forms of metabolic energy, just as enzymatic reactions can be coupled to the hydrolysis or synthesis of ATP. For example, molecules can be transported in an energetically unfavorable direction across a membrane (e.g., against a concentration gradient) if their transport in that direction is coupled to ATP hydrolysis as a source of energy—a process called active transports as shown in figure below. The free energy stored as ATP can thus be used to control the internal composition of the cell, as well as to drive the biosynthesis of cell constituents.

Model of active transport Energy derived from the hydrolysis of ATP is used to transport H+ against the electrochemical gradient (from low to high H+ concentration). Binding of H^+ is accompanied by phosphorylation of the carrier protein, which induces a conformational change that drives H^+ transport against the electrochemical gradient. Release of H^+ and hydrolysis of the bound phosphate group then restore the carrier to its original conformation.

Cytoplasm

In a eukaryotic cell, the nucleus is surrounded by the cytoplasm, which in animal cells usually accounts for nine-tenths of the cell's volume; the cytoplasm is surrounded in turn by a cell membrane, or plasmalemma. The cytoplasm contains within a protein polymer matrix several types of minute, functionally specialized cell organs, or organelles, most of them present in multiple copies. In prokaryotic cells (bacteria), by contrast, the genome is not separated from the cytoplasm, there are no membrane-bounded organelles, and the plasma membrane subserves many of the specialized functions of eukaryotic organelles.

Prokaryotic and Eukaryotic Cells use Related Strategies to Compartmentalize their Biosynthetic Reactions

Prokaryotic cells have only one membrane, located at the cell surface (although some bacteria have a double membrane). In contrast, eukaryotic cells, such as those of animals and plants, have, in addition to their plasmalemma, at least a dozen different types of chemically specific membranes that create separate intracellular compartments with different microenvironments required by various processes, such as respiration, photosynthesis, protein synthesis, and intracellular digestion.

In prokaryotic cells, the plasmalemma is the site of a number of important biosynthetic activities, including the synthesis of membrane lipids, adenosine triphosphate, and cell-wall components. This membrane also contains a multitude of transporters, receptors, and sensors for chemotactic movements, which create the interface of the bacterium with its extracellular environment.

In eukaryotic cells, comparable activities are differently distributed. Some of them (such as those of the transporters and receptors) remain in the plasmalemma, but others are relocated to different intracellular membrane systems. Lipid synthesis, for instance, in animal cells occurs only in the endoplasmic reticulum (ER), a network of membrane-bounded channels that pervades the cytoplasm. In plant cells, lipid synthesis occurs in plastids. From the ER or plastids, newly synthesized lipids are distributed to all other cellular membranes. The ER also contains a complex set of enzymes that modify lipid-soluble aromatic compounds imported by the cell from the environment. The modifications increase the water solubility of these compounds, thereby facilitating their elimination. Since sterols, drugs, herbicides, toxins, and chemical carcinogens are among these compounds, the relevant enzymes constitute an intracellular detoxifying system.

The ER membrane is also the site of important steps in protein traffic regulation. Issues awaiting resolution include the means by which proteins and lipids are transported from their site of synthesis in the ER to a multiplicity of destinations and the way differences in lipid chemistry are established and maintained in different membranes. We must also advance our understanding of the ER detoxifying system to shed light on problems related to chemically induced cancers and toxic effects of chemical pollutants in the environment.

Protein Synthesis and Regulation

Protein Synthesis

For all cells, protein synthesis is a major, continuous activity needed for the production of intracellular enzymes, contractile and cytoskeletal assemblies, membranes, ribosomes, chromosomes, and many other functionally important macromolecular assemblies. In more complex multicellular organisms, it is also needed for the production of proteins destined for export out of the cell, such as enzymes, hormones, growth factors, blood proteins, antibodies, or components of the extracellular matrix.

In all cells, proteins are synthesized on ribosomes, which translate into amino acid sequences the instructions received from active genes in the form of mRNAs. The ribosomes themselves are macromolecular assemblies of ribosomal RNA and protein molecules. The two sets of components are produced separately in the cytoplasm (proteins) and in the nucleus (ribosomal RNAs) and are modified and assembled in a special compartment in the nucleus (the nucleolus) before being transferred to the cytoplasm as part of a functioning ribosome.

In both prokaryotic and eukaryotic cells, ribosomes are basically similar and protein synthesis proceeds by similar steps. In eukaryotic cells, however, the ribosomes are larger and require more factors for their activity. These differences are probably related to the emergence of more versatile regulatory processes in eukaryotes.

Protein Traffic Control

In eukaryotic cells, ribosomes and protein synthesis occur primarily in the cytosol, to which mRNAs have direct access from the nucleus and in which the pool of amino acids and all ancillary factors required for protein synthesis are located. Only 2 percent of the total protein production is accounted for by small mitochondrial ribosomes whose products are used exclusively in mitochondria. In plant cells a somewhat larger fraction of the cell's protein is produced by chloroplasts (or other differentiated forms of plastids). From the cytosol, proteins are accurately directed to more than 20 different sites of final functional residence. These sites are membranes or compartment contents, each endowed with chemical specificity.

ER Targeting System

Many proteins are directed to the ER membrane as they are being synthesized. The selectivity is based on mutual recognition between signals (called signal sequences) within the amino acid sequence of the protein to be transported and a signal recognition complex. This complex involves a ribonucleoprotein particle (called a signal recognition particle) and at least one transmembrane protein (its receptor) in the target ER membrane. The ER targeting system recognizes and processes proteins destined for secretion and lysosomes (storage compartments for degradative enzymes), as well as membrane proteins for many intracellular compartments. Secretory and lysosomal proteins are fully translocated across the ER membrane into the ER cisternal space. Membrane proteins are partially translocated and remain anchored in the membrane. Recent experiments indicate that a membrane protein can be converted into a secreted polypeptide, and conversely, a secretory protein can be convened into a membrane protein by deleting or adding the information for the membrane anchoring sequences from their mRNAs.

The mode of operation of the ER targeting system has been elucidated in reconstituted in vitro systems in which ribosomes are allowed to translate into proteins the genetic information encoded in specific mRNAs in the presence or absence of ER-membrane vesicles in vitro. The results show that many components of the systems are functionally equivalent in different species, phyla, and even kingdoms, which implies that this part of the traffic regulation system originated early in evolution and has been conserved.

Post-ER Steps in Traffic Control

Once past the ER, proteins are moved within the cell through a specialized membrane-bounded compartment called the Golgi complex, where they are further modified by glycosylation, sulfation, and proteolysis; they are sorted while in transit to lysosomes, secretion vacuoles, or different membranes. The Golgi complex itself contains at least three subcompartments. Transport from the ER to Golgi subcompartments, from one subcompartment to another in the Golgi complex, and finally, from the last Golgi elements to the plasmalemma requires energy and is affected by vesicular carriers; thus, past the ER, protein traffic can be regulated at least in part by controlling the movements of vesicular carriers. These carriers apparently recycle continuously among compartments. The best known among these vesicular carriers are the secretion granules or secretion vacuoles of various glandular cells. They transport products to the cell surface and discharge them into the extracellular medium by a process known as exocytosis. Sorting of the proteins to their ultimate destinations probably involves mutual recognition between a signal and a receptor, but so far only the signal for lysosome proteins has been chemically defined. Its receptor has been isolated and partially characterized and its gene sequenced. Reactants involved in the sorting of other proteins remain unknown, as are the signals and receptors that regulate the traffic of vesicular carriers. Studies to identify them are being actively pursued.

Other Traffic-control Systems

The ER targeting system (which includes the Golgi complex) is undoubtedly the most complex component of the overall protein traffic-control system of the cell. The other components are simpler, and most of them probably transport the protein in a single step: from the cytosol directly to the final destination. The amino acid sequence of the signal that directs certain proteins to the nucleus is known in a few cases, but the corresponding receptor remains to be identified. A substantial body of information already exists about traffic regulation of proteins targeted to mitochondrial and chloroplast membranes. Among the protein products of plant nuclear genes are some that function in the mitochondria and some in chloroplasts; it is not known how the systems differ.

In a simpler form, protein traffic regulation occurs in prokaryotic cells and has been studied extensively in gram-negative bacteria, which are provided with two concentric membranes at the cell surface. The number of final destinations is considerably fewer—the two membranes and the intervening space and perhaps the outside of the cell. A signal, generally comparable to that found in eukaryotic proteins targeted to the ER membrane, has been identified and analyzed in detail by sequencing and by extensive genetic modifications. This line of work has led to the recognition of functionally critical residues in the signal sequence, but the other components of the system are still unknown.

We can anticipate considerable activity in this fertile and exciting area, especially in relation to the identification and characterization of signals and their receptor partners and to the intracellular location of receptors and sorters (molecules that control the selection and movement of proteins from one compartment to another). The reasons for removing the signal sequence are not clear, nor are the reasons for the complexity of the enzyme that effects the removal. The enzyme may have additional functions since it consists of six different proteins. The translocation process itself is still a mystery.

Structural biology is likely to yield three-dimensional information on such interactions if large enough quantities of relevant proteins can be obtained. The main goal is to understand how cells process their many proteins and how they achieve and maintain the chemical specificity of their membranes. Another basic process that remains to be understood in molecular terms is membrane fusion. The process is critical for cell division, cell fusion in egg fertilization, and vesicle fusion along different pathways of intracellular transport. Membrane fluidity is a prerequisite for membrane fusion. It is also a prerequisite for membrane growth, which appears to proceed by expanding preexisting membranes. At the same time, intact diffusion barriers need to be maintained in highly dynamic membrane systems.

Protein Synthesis and Regulation

Transcription and RNA processing are followed by translation, the synthesis of proteins as directed by mRNA templates. Proteins are the active players in most cell processes, implementing the myriad tasks that are directed by the information encoded in genomic DNA. Protein synthesis is thus the final stage of gene expression. However, the translation of mRNA is only the first step in the formation of a functional protein. The polypeptide chain must then fold into the appropriate three-dimensional conformation and, frequently, undergo various processing steps before being converted to its active form. These processing steps, particularly in eukaryotes, are intimately related to the sorting and transport of different proteins to their appropriate destinations within the cell.

Although the expression of most genes is regulated primarily at the level of transcription, gene expression can also be controlled at the level of translation, and this control is an important element of gene regulation in both prokaryotic and eukaryotic cells. Of even broader significance, however, are the mechanisms that control the activities of proteins within cells. Once synthesized, most proteins can be regulated in response to extracellular signals by either covalent modifications or by association with other molecules. In addition, the levels of proteins within cells can be controlled by differential rates of protein degradation.

Mitochondria

Mitochondria are rod-shaped organelles that can be considered the power generators of the cell, converting oxygen and nutrients into adenosine triphosphate (ATP). ATP is the chemical energy "currency" of the cell that powers the cell's metabolic activities. This process is called aerobic respiration and is the reason animals breathe oxygen. Without mitochondria (singular, mitochondrion), higher animals would likely not exist because their cells would only be able to obtain energy from anaerobic respiration (in the absence of oxygen), a process much less efficient than aerobic respiration. In fact, mitochondria enable cells to produce 15 times more ATP than they could otherwise, and complex animals, like humans, need large amounts of energy in order to survive.

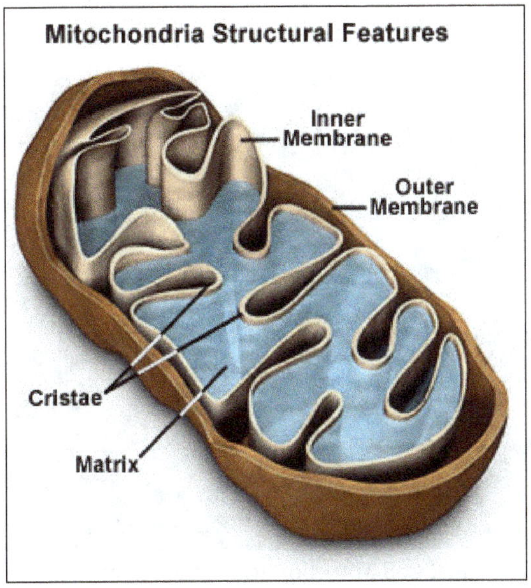

Mitochondria Structural Features.

The number of mitochondria present in a cell depends upon the metabolic requirements of that cell, and may range from a single large mitochondrion to thousands of the organelles. Mitochondria, which are found in nearly all eukaryotes, including plants, animals, fungi, and protists, are large enough to be observed with a light microscope and were first discovered in the 1800s. The name of the organelles was coined to reflect the way they looked to the first scientists to observe them, stemming from the Greek words for "thread" and "granule." For many years after their discovery, mitochondria were commonly believed to transmit hereditary information. It was not until the mid-1950s when a method for isolating the organelles intact was developed that the modern understanding of mitochondrial function was worked out.

The elaborate structure of a mitochondrion is very important to the functioning of the organelle as shown in figure above. Two specialized membranes encircle each mitochondrion present in a cell, dividing the organelle into a narrow intermembrane space and a much larger internal matrix, each of which contains highly specialized proteins. The outer membrane of a mitochondrion contains many channels formed by the protein poring and acts like a sieve, filtering out molecules that are too big. Similarly, the inner membrane, which is highly convoluted so that a large number of infoldings called cristae are formed, also allows only certain molecules to pass through it and is much more selective than the outer membrane. To make certain that only those materials essential to the matrix are allowed into it, the inner membrane utilizes a group of transport proteins that will only transport the correct molecules. Together, the various compartments of a mitochondrion are able to work in harmony to generate ATP in a complex multi-step process.

Mitochondria are generally oblong organelles, which range in size between 1 and 10 micrometers in length, and occur in numbers that directly correlate with the cell's level of metabolic activity. The organelles are quite flexible, however, and time-lapse studies of living cells have demonstrated that mitochondria change shape rapidly and move about in the cell almost constantly. Movements of the organelles appear to be linked in some way to the microtubules present in the cell, and are probably transported along the network with motor proteins. Consequently, mitochondria may be organized into lengthy traveling chains, packed tightly into relatively stable groups, or appear in many other formations based upon the particular needs of the cell and the characteristics of its micro tubular network.

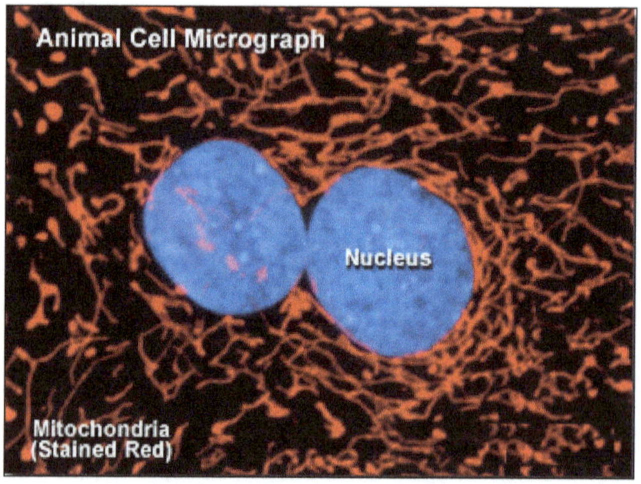

Animal Cell Mitochondria in the Microscope.

The figure above is a digital image of the mitochondrial network found in the ovarian tissue from a mountain goat relative, known as the Himalayan Tahr, as seen through a fluorescence optical microscope. The extensive intertwined network is labeled with a synthetic dye named MitoTracker Red (red fluorescence) that localizes in the respiring mitochondria of living cells in culture. The rare twin nuclei in this cell were counterstained with a blue dye (cyan fluorescence) to denote their centralized location in relation to the mitochondrial network. Fluorescence microscopy is an important tool that scientists use to examine the structure and function of internal cellular organelles.

The mitochondrion is different from most other organelles because it has its own circular DNA (similar to the DNA of prokaryotes) and reproduces independently of the cell in which it is found; an apparent case of endosymbiosis. Scientists hypothesize that millions of years ago small, free-living prokaryotes were engulfed, but not consumed, by larger prokaryotes, perhaps because they were able to resist the digestive enzymes of the host organism. The two organisms developed a symbiotic relationship over time, the larger organism providing the smaller with ample nutrients and the smaller organism providing ATP molecules to the larger one. Eventually, according to this view, the larger organism developed into the eukaryotic cell and the smaller organism into the mitochondrion.

Mitochondrial DNA is localized to the matrix, which also contains a host of enzymes, as well as ribosomes for protein synthesis. Many of the critical metabolic steps of cellular respiration are catalyzed by enzymes that are able to diffuse through the mitochondrial matrix. The other proteins involved in respiration, including the enzyme that generates ATP, are embedded within the mitochondrial inner membrane. Infolding of the cristae dramatically increases the surface area available for hosting the enzymes responsible for cellular respiration.

Mitochondria are similar to plant chloroplasts in that both organelles are able to produce energy and metabolites that are required by the host cell. Mitochondria are the sites of respiration, and generate chemical energy in the form of ATP by metabolizing sugars, fats, and other chemical fuels with the assistance of molecular oxygen. Chloroplasts, in contrast, are found only in plants and algae, and are the primary sites of photosynthesis. These organelles work in a different manner to convert energy from the sun into the biosynthesis of required organic nutrients using carbon dioxide and water. Like mitochondria, chloroplasts also contain their own DNA and are able to grow and reproduce independently within the cell.

In most animal species, mitochondria appear to be primarily inherited through the maternal lineage, though some recent evidence suggests that in rare instances mitochondria may also be inherited via a paternal route. Typically, a sperm carries mitochondria in its tail as an energy source for its long journey to the egg. When the sperm attaches to the egg during fertilization, the tail falls off. Consequently, the only mitochondria the new organism usually gets are from the egg its mother provided. Therefore, unlike nuclear DNA, mitochondrial DNA doesn't get shuffled every generation, so it is presumed to change at a slower rate, which is useful for the study of human evolution. Mitochondrial DNA is also used in forensic science as a tool for identifying corpses or body parts, and has been implicated in a number of genetic diseases, such as Alzheimer's disease and diabetes.

The mitochondrion is characterized as the power plant of the cell, because it performs the enzyme-catalyzed, stepwise oxidation of nutrients (such as sugars, fats, and amino acids) in a process called respiration. The most interesting product of this process is ATP, the direct source of the energy required for most of the chemical work the cell must perform to power its growth, movement, synthesis of new components, and other functions. A relatively small amount of ATP is produced in the cytosol during sugar catabolism (glycolysis), but by far the largest amount is generated in mitochondria. The mechanism of mitochondrial ATP synthesis has stubbornly resisted full elucidation, but progress continues to be made as a result of our insistent probing into this fundamental energy-transducing process.

Mitochondria are also of interest because they contain their own complement of DNA, which cooperates with the DNA of the nucleus in the control of mitochondrial formation. The origin and evolution of the mitochondrion are linked to the origin and evolution of eukaryotic cells. Understanding of mitochondrial function in turn sheds light on a wide array of fundamental and practical issues, ranging from certain metabolic and genetic diseases to evolution itself.

New Mitochondria Arise from Existing Ones and they are Characterized by Unique Functions

Mitochondria have two membranes (inner and outer) that define two concentric separate spaces. The inner space houses hundreds of enzymes, including those involved in the oxidative reactions that supply the energy needed for cell function. The inner mitochondrial membrane contains the energy-conversion apparatus. The majority of the mitochondrial proteins are specified by nuclear genes; they are synthesized on ribosomes in the cytoplasm and then imported into the organelle. A limited set of proteins of the inner mitochondrial membrane, namely, some protein subunits of the oxidative phosphorylation apparatus, are encoded in DNA molecules located within the organelle itself and are synthesized by an organelle-specific translation system. The distinctive structural RNA components of this system—RNAs and transfer RNAs—are also encoded in mitochondrial DNA. Mitochondria do not arise de novo in the cell by self-assembly of their constituent molecules, but are formed by growth and division of existing mitochondria.

The mitochondrial DNA from several organisms, including humans, has been completely sequenced, and much of its informational content has been elucidated. Furthermore, all mitochondrial gene products in humans have been functionally identified. A dramatic outcome of these studies has been the discovery that the mitochondrial genetic system in the organisms studied, except plants, uses a code that differs in several respects from the universal code and, in addition, utilizes for reading the code a novel mechanism, which requires only a restricted set of transfer RNAs.

Excellent Opportunities Exist to Study the Mechanisms of Expression and Replication of Mitochondrial DNA

Studies of the enzymes and ancillary proteins responsible for mitochondrial DNA replication, DNA transcription into RNA, and RNA processing to mature RNA species are making rapid progress, aided by the development of soluble in vitro preparations derived from broken mitochondria and by the use of recombinant DNA technologies. Specific proteins have already been identified and, in some cases, isolated in partially or completely pure form.

As in the case of many nuclear gene transcripts, the coding sequences of several mitochondrial gene transcripts in lower eukaryotes especially yeast and filamentous fungi are interrupted by nonfunctional segments, or introns. These introns must be removed to produce the mature RNA. The transcripts of some mitochondrial genes have the remarkable capacity to excise their own introns in vitro in the absence of protein; that is, they function as enzymes acting on themselves. The discovery of mitochondrial introns has opened an active field of research, which is expected to provide deep insights into the mechanisms of RNA splicing in general and into the origin and evolution of introns.

The Formation of New Mitochondria is under the Control of Both the Nucleus and Mitochondria

The dual control of mitochondrial formation is most dramatically illustrated by the chimeric structure of nearly all the enzyme complexes of the oxidative phosphorylation system. Each such complex contains some subunits encoded in the nucleus and some encoded in mitochondrial DNA. Because of this dual control, the formation of functional mitochondria requires a quantitative and temporal coordination of expression of the relevant nuclear and mitochondrial genes.

Two main classes of nuclear genes are the object of intensive investigation based on recombinant DNA techniques and on structural and functional approaches:

- Genes coding for subunits of the enzyme complexes of the oxidative phosphorylation system and for mitochondrial carriers used in metabolite transport.
- Genes coding for proteins involved in the expression and replication of the mitochondrial genome.

Most of the latter genes are probably distinct from those that code for the homologous components of the nuclear-cytoplasmic apparatus, although interesting exceptions have recently been reported. They concern the possible existence of common nuclear genes for cytoplasmic and mitochondrial components, which could account for at least some of the reported influences of the mitochondrial genome on the remainder of the cell.

Proteins are Imported into Mitochondria

The hundreds of distinct polypeptides that make up a mitochondrion are distributed in a specific way in four compartments: the outer membrane, inter-membrane space, the inner membrane, and inner mitochondrial space. After their synthesis on cytoplasmic ribosomes, the nuclear gene-coded polypeptides are imported to their correct location within the mitochondrion. Biochemical studies and the application of recombinant DNA technology have shown that proteins destined for one of the three innermost compartments are usually made from precursors with extensions at the amino-terminal end; these extensions can be as long as 100 amino acids and function as signals directing the proteins to the proper location.

Crystallographic Studies Should Reveal the Tertiary Structure of the Oxidative Phosphorylation Apparatus

The subunit composition of the enzyme complexes of the oxidative phosphorylation system is largely known, as is the location—nuclear or mitochondrial—of the genes specifying these subunits. From the nucleotide sequence of these genes, the amino acid sequence of the subunits can be inferred. This information, together with data derived from other approaches to studies of the spatial relations of the subunits in each enzyme complex, is likely to provide useful models of the three-dimensional structure of each complex and of its topology in the inner mitochondrial membrane. These models will help in interpreting the results of ongoing functional studies on the individual complexes in intact mitochondria and in reconstructed systems.

Nucleus

The nucleus is a membrane bound organelle found in the majority of eukaryotic cells. It is the largest organelle of the eukaryotic cell, accounting for around 10% of its volume. It houses the genome, and through gene expression, it co-ordinates the activities of the cell.

Structure

The nucleus is a relatively large and spherical membrane-bound organelle. The nucleus itself is comprised of distinct components, and understanding their structure allows a deeper understanding of their function.

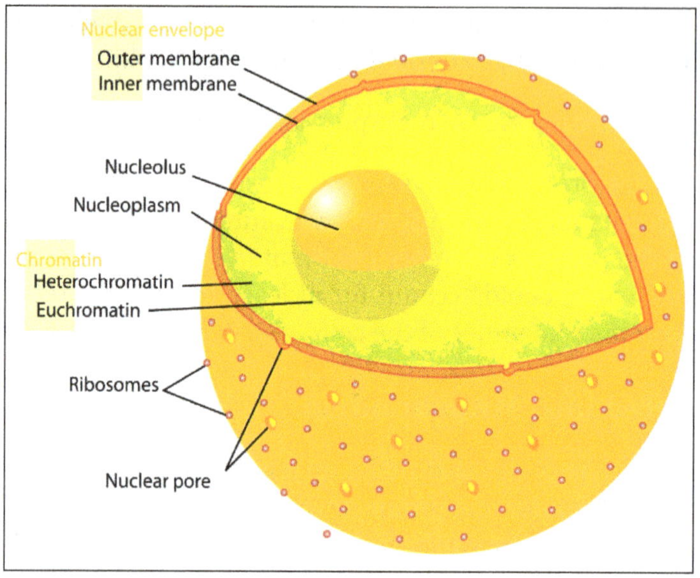

The nucleus.

Nuclear Envelope

The nucleus is completely surrounded by the nuclear envelope. This consists of both an inner and outer membrane which run parallel to each other. The envelope is perforated by small gaps known as the nuclear pores. These pores are around 100nm wide in true diameter, however due to the presence of central regulatory proteins the true size of the gap is around 9nm.

This small size controls the passage of molecules into and out of the nucleus. Larger molecules such as larger proteins and nucleic acid are unable to pass through these pores, and so the function of the nuclear envelope is to selectively separate the contents of the nucleus from that of the cytoplasm.

Nuclear Lamina

Mechanical support for the nucleus is provided by the nuclear lamina. This is a protein mesh, which is more organised on the internal surface on the nucleus than on the cytoplasmic surface.

Chromatin

Chromatin describes DNA that is complexed with proteins. The primary protein components of chromatin are histones, which are highly basic proteins that associate readily with DNA. Histones combined with DNA form nucleosomes, which are the subunit of chromatin. Specifically, a nucleosome describes a segment of DNA associated with 8 histone proteins. By associating with histones, DNA is more compact and able to fit into the nucleus.

Chromatin can exist as either euchromatin or heterochromatin. Euchromatin is the form of chromatin present during gene expression, and has a characteristic 'beads on a string' appearance. It is activated by acetylation. In contrast, heterochromatin is the 'inactive' form, and is densely packed. On electron microscopy, euchromatin stains lighter than heterochromatin which reflects their relative densities.

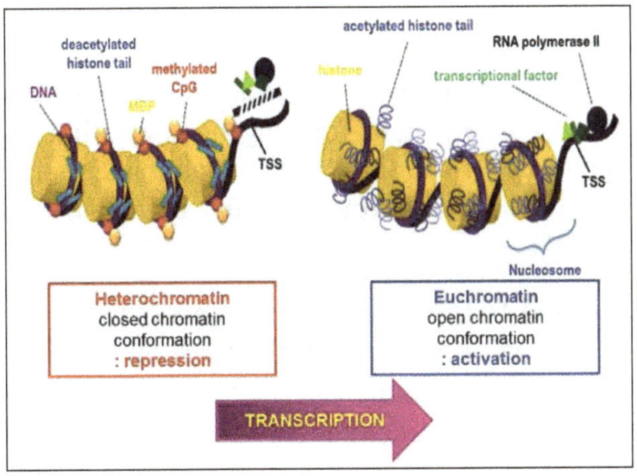

Schematic diagram of euchromatin and heterochromatin.

Nucleolus

The nucleolus is the site of ribosome and ribosomal RNA production. On microscopy, it appears as a large dense spot within the nucleus. After a cell divides, a nucleolus is formed when chromosomes are brought together into nucleolar organizing regions. During cell division, the nucleolus disappears.

Function

- Cell compartmentalization: The presence of a selectively permeable nuclear envelope separates the contents of the nucleus from that of the cytoplasm.
- Gene expression: Gene expression first requires transcription, which is the process by which DNA is transcribed into mRNA. As the nucleus is the site of transcription, proteins within the nucleus play a key role in regulating the process.
- Processing of pre-mRNA: Newly synthesized mRNA molecules are known as pre-mRNA. Before they exit the nucleus, they undergo a process known as post-transcriptional modification where molecules are added or removed from the structure.

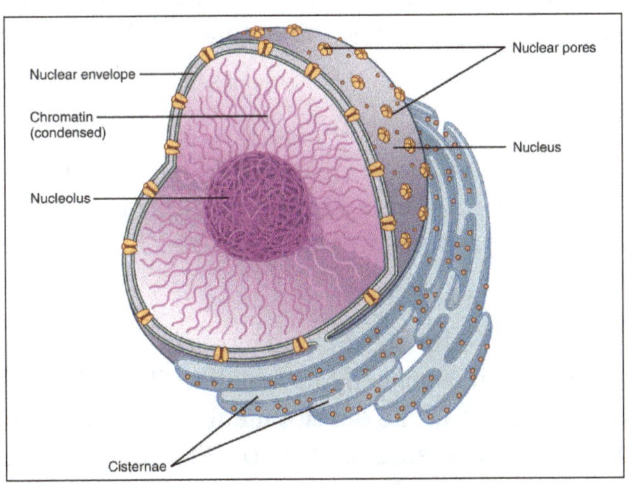

Nucleus with the cisternae of a continuous endoplasmic reticulum highlighting its main features.

Chromosomes

Chromosomes are a complex network of protein and DNA helps DNA to organize during cell division and protects it. In 1842, Karl Wilhelm von Nageli, a swiss botanist discovered a structure which later known as chromosomes. In 1882, Wilhelm Hofmeister denoted the term as "chromosomes". Later on in the year, 1883 Wilhem roux speculated chromosomes as a carrier of genetic information.

Unlike the prokaryotic DNA, the DNA of eukaryotes is arranged on chromosomes having centromere, telomere, p arm and q arm. A prokaryotic chromosome is either made up of DNA or RNA (RNA in case of some viruses)- but not located in the nucleus. Different species have a different number of chromosomes based on their genome size. The number of chromosomes in different organisms are given below:

Organism	Number of chromosomes (2n)
Human	46
Potato	48
RIce	24
Monkey	42
Mouse	40

- "DNA of eukaryotes such as plants and animals is arranged as a tightly packed thread-like structure known as a chromosome."
- "A complex network of DNA and protein- coiled around each other and helps to fit DNA inside the nucleus is known as a chromosome."

Human Chromosomes

46 chromosomes are present in each human cell, except germ cells. All chromosomes are present in a pair thus a total of 46 chromosomes are present in 23 pairs called a diploid number of chromosomes. All the somatic cells have a diploid number of chromosomes while egg and sperm germ cells have a haploid number of chromosomes. All the nuclear DNA of a cell is arranged on these 23 pairs of chromosomes, however, some linear or circular DNA is also present in the cytoplasm.

X and Y Chromosome

The 22 pairs of chromosomes are the same in all-male and females, however, one pair of sex chromosomes decide their gender. A pair of the X chromosome is present in females while a single X along with a short Y chromosome is present in males. Hence XX in females and XY in the male are denoted as sex chromosomes.

Interestingly, a mechanism called X-chromosome inactivation helps to inactivate one X chromosome for regulating gene expression, especially in females. The SRY a sex-determining region of Y, responsible for maleness and helps in the development of male phenotypes although it is solely not decided the fate of embryo. The presence or absence of the X chromosome decides whether the developing embryo becomes female or male.

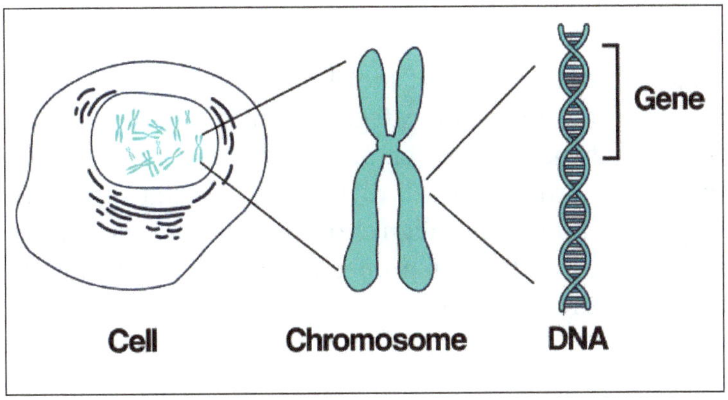

The location of genes on a chromosome.

Structure of the Chromosome

- How DNA is arranged on the chromosome is a complex process. DNA is a double-stranded molecule and helical in shape.
- Imagine a rope and arrangement of threads in a rope. This spiral arrangement creates tension on the remaining strand of DNA. Now the DNA wraps on each other and creates a supercoiled structure.
- It binds with Histone protein molecules (H1, H2A, H2B, H3 and H4) and forms a nucleosome.
- It looks like a solenoid that has the beads-on-string like structure and, later turned into long chromatin. Finally, the chromatid formed from chromatin is attached to the centromere and formulates the chromosome.

Centromere and arms are the main components of a chromosome:

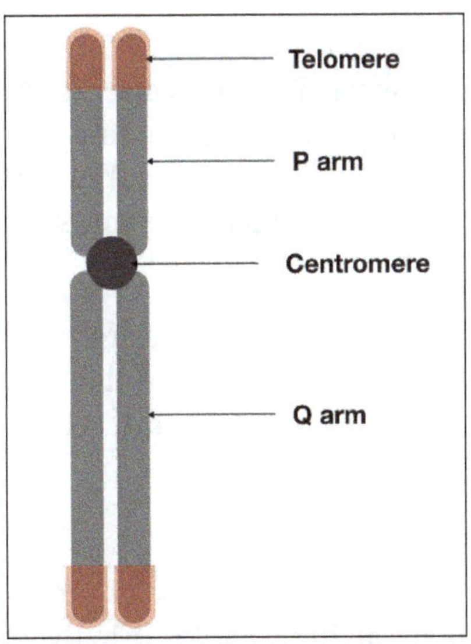

The graphical representation of the structure of chromosome- centromere, arms and telomeres.

Centromere

First of all, the centromere is not a center of the chromosome, remember that, based on the location of the centromere, chromosomes are categorized into different categories. Actually, a constricted, narrow and somewhat rounded region of a chromosome which holds chromatids known as the centromere. During the process of cell division, and when replication is in progress, the centromere plays an important function to align chromosomes properly to replicate. It provides an attachment point to complete the replication properly. Notably, important genes are not present in the centromeric region of the chromosome.

Arms

The "p arm" and "q arm" of the chromosome are attached to the centromeres. Although, based on the location of the centromere, the length of the arms varies. For example, the p and q arm of the metacentric chromosome 1 has almost the same length while the p arm of acrocentric chromosome 21 is very short. The arms are, we can say the main body of the chromosomes, having all the essential genes on it.

Arms are the complex network of protein and DNA where genes are located. The densely packed area of arms is called heterochromatin region- a gene less region, rich in non-coding DNA. On the other hand, the loosely packed region is known as the euchromatin region- a gene-rich region. The tip of each arm is protected by the structure called telomeres.

Telomeres

Telomeres are the end of chromosomes that protects gene-rich region- and are made up of the repetitive DNA sequence, the non-coding DNA sequences on the telomeres are categorized in microsatellite and minisatellite. The repetitive sequences are highly packed and thus do not encodes any protein however, it protects other genes from the cell's own mistake called "end replication problem".

The end replication problem occurs when DNA polymerase starts incorporating wrong nucleotides, especially at the end of the replication. Protein-coding genes are located on arms thus replicates properly. But once the polymerase reaches to end, the replication stops or incorporates the wrong bases henceforth after each round of replication some sequences from the telomeres can not be replicated and lost.

After each round of replication, the length of telomeres reduces, once it disappears cells will die and consequently the organism. Therefore it is believed that the length of the telomere is directly proportional to the age of the organism. Contrary to this, if telomere shortening not happened, it can cause cancer.

Classification of Chromosomes

The centromere of chromosomes plays an important role in the characterization of it. Based on the location of the centromere, the 23 pairs of human chromosomes can be divided into 7 categories.

Group	Size	Position of centromere	Number of chromosomes
Group A	Large	Metacentric	1, 2 and 3
Group B	Large	Submetacentric	4 and 5
Group C	Medium	Submetacentric	6, 7, 8, 9, 10, 11,12 and X
Group D	Medium	Acrocentric	13, 14 and 15
Group E	Relatively short	Submetacentric	16, 17 and 18
Group F	Short	Meta or submeta	19 and 20
Group G	Short	Acrocentric	21, 22 and Y

- Metacentric Chromosomes: In metacentric chromosomes, a centromere is located exactly or nearly exactly in the center of both p and q arms. Due to this p and q arms are equal in length.
- Submetacentric Chromosomes: In this type of chromosomes, a centromere is shifted a little towards one of the two arms. Thus the length of the p and q arm are almost similar but not equal.
- Acrocentric Chromosomes: In the acrocentric chromosomes, the centromere located very close to the p arm and therefore the p arm is very shorter than the q arm. The q arm is longer than the p arm but the p arm is a little larger than the telocentric chromosomes.
- Telocentric Chromosomes: In this type of chromosomes, the centromere is located very close to the p arm and thus the p arm is barely visible or absent. However, the q arm is still long enough to distinguish. Notably, humans do not have telocentric chromosomes while all the chromosomes of the house mouse are telocentric.

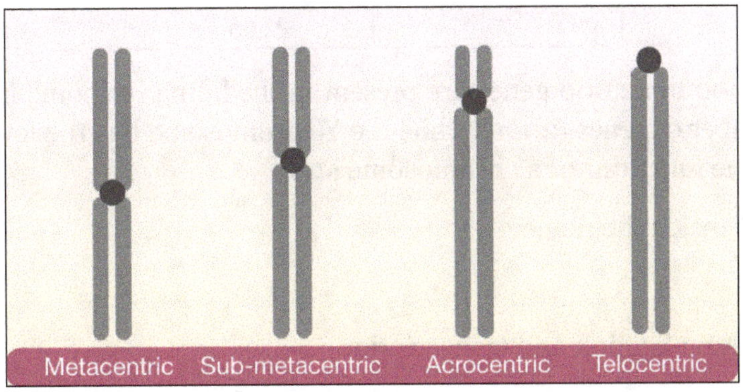

Classification of chromosomes based on the location of centromere.

The Function of Chromosomes

Chromosomes facilitate proper cell division and replication. The main function of the chromosome is to fit the DNA inside the nucleus. Hence it is very important to fit it inside the nucleus which is facilitated by chromosomes. By interacting with proteins DNA forms a coiled structure- chromosome.

Chromosomes also help in inheriting genes or DNA from parents to their offsprings. Furthermore, sex chromosomes decide the sex of the embryo. The process of sex determination and sex differentiation is governed by genes located on autosomes and sex chromosomes.

Table: Size and number of genes on different human chromosomes.

Chromosome	Number of genes
1	2000
2	1300
3	1000
4	1000
5	900
6	1000
7	900
8	700
9	800
10	700
11	1300
12	1100
13	300
14	800
15	600
16	800
17	1200
18	200
19	1500
20	500
21	200
22	500
X	800
Y	200

A total of around 21,000 to 22,000 genes are present in the human genome in which chromosome 1 has the highest number of genes (2000) while the Y chromosome has the lowest number of genes (200). Conclusively the functions of a chromosome are:

- Make DNA fit inside the nucleus.
- Protect genes.
- Facilitate DNA replication and transcription.
- Store and transfer genetic information.
- Protect DNA from damage.
- Helps in regulating gene expression.

Chromosomal Abnormalities

Chromosomal abnormalities occur due to the change in either chromosome number or chromosome structure.

- Deletion, duplication, translocation, inversion and addition are some of the structural chromosomal abnormalities while trisomy, tetrasomy and monosomy are numerical chromosomal abnormalities.

- Interestingly: In numerical chromosomal abnormalities change in the total number of chromosomes results in an abnormal genetic condition.
- Trisomy: One extra chromosome is added into the genome. For example trisomy 21, in which three different 21st chromosomes occur. This condition results in genetic abnormality called down syndrome- a type of mental disorder. Another example of trisomy is trisomy 18 called Edwards syndrome.
- Monosomy: One chromosome is absent from the pair, the condition is called monosomy. For example, monosomy of X often known as XO or turner syndrome which is a type of complex genetic disorder.
- Uniparental disomy: As we know that one chromosome from father and one chromosome from mother comes in offspring. But in uniparental disomy, both chromosomes come from the same parents. Although the total number of chromosomes remains unchanged, it causes genetic abnormality. For example Prader-Willi syndrome.
- Anyway: One of the classical examples, of structural chromosomal abnormality, is the translocation between chromosome 9 and 22 which is called as Philadelphia chromosome.

Chromosome Analysis

Alike DNA mutations, chromosomal abnormalities are also common in humans. Thus for identifying chromosomal abnormality, we need techniques for chromosomal analysis. In recent days two of the best chromosomal analysis methods- karyotyping and chromosomal microarray are used routinely.

Karyotyping

Karyotyping- a cytogenetic method is often known as PBLC- peripheral blood leukocyte culture is employed to encounter numerical and lager structural chromosomal abnormalities. In this method, the metaphase cells are harvested for getting metaphase chromosomes and stained using the Giemsa stains.

A traditional, conventional and widely used method karyotyping is used since long for the evaluation of chromosomal abnormalities. It is also used for the analysis of chromosomes in other species as well. The major limitation of karyotyping is, it cannot identify minor deletions or duplications or other structural abnormalities because of the lower resolution banding.

Chromosome Microarray

Chromosome microarray is a DNA hybridization-based method practiced for detecting many structural chromosomal aberrations in a single assay. In the chromosome microarray often known as "whole chromosome microarray," thousands of different probes are immobilized on the solid glass surface. Though it is called chromosome microarray we are using total genomic DNA, instead of metaphase chromosomes.

The microarray detector detects abnormality if any, arrange it accordingly on each chromosome. The main advantage of the present technology is that it is robust and automated as compared with conventional karyotyping. Besides this, we can screen thousands of different copy number variations at once. The results of karyotyping and chromosome microarray are shown in the figure below.

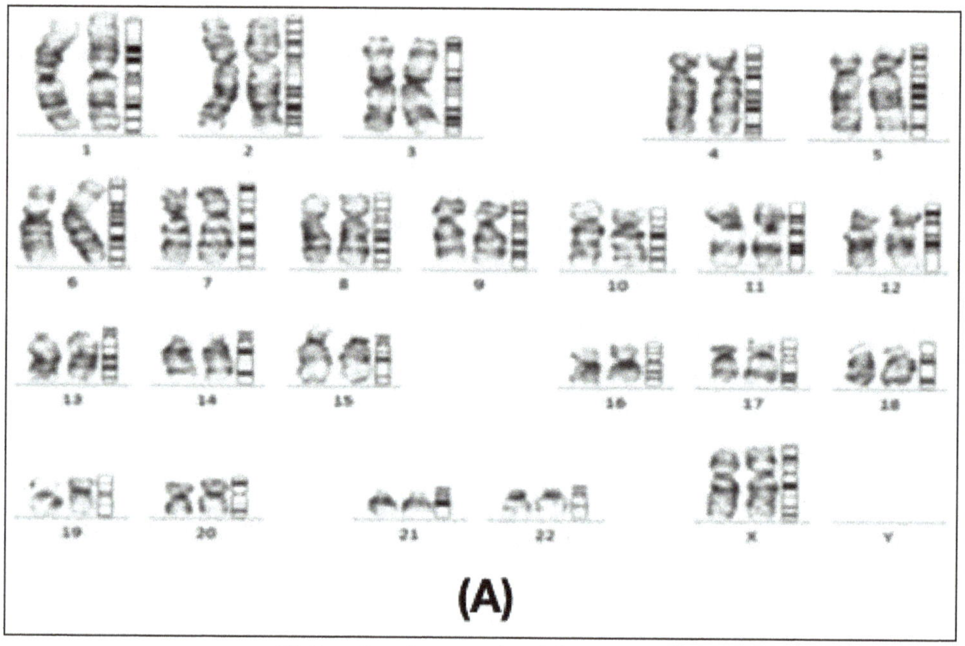

(A) Results of conventional karyotyping.

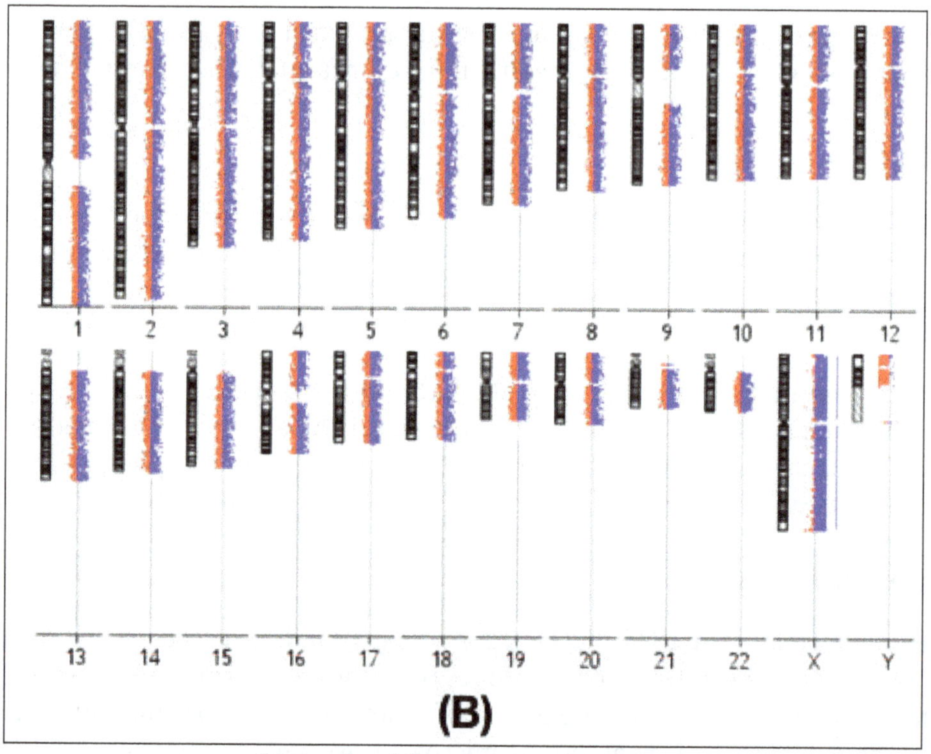

(B) Results of Chromosome microarray.

Gene Action

The functioning of genes (hereditary units) in determining the structural and functional characteristics of an individual, that is, its phenotype. Gene action is studied by two somewhat different, but complementary, approaches the analysis of changes which occur in the phenotype when a gene mutates, or is changed in dosage, or in position relative to other genes; this is frequently called the study of phenogenetics; and the more direct approach, which attempts to determine the actual means by which genes exert their control over metabolism and the processes of development in multicellular, differentiated organisms. The more direct approach is best described as study of primary gene action, but includes study of the interaction of primary or secondary products of gene action.

All genes, with the exception of those in ribonucleic acid (RNA) viruses, are constituted of deoxyribonucleic acid (DNA), and the primary action of the great majority of them is to initiate a series of events leading directly or indirectly to the determination of the amino acid sequences of specific polypeptides.

The base sequence of one of the chains of the DNA double helix constituting the gene is transcribed into an RNA molecule with a chain of complementary bases in the presence of RNA polymerases. This RNA molecule may then frequently become a messenger RNA (mRNA) by some alteration of the original transcript, or it may become a transfer RNA (tRNA) or a ribosomal RNA (rRNA).

The primary action of a gene is transcription, but the expression of this action lies in the next step—translation—for many genes. The mRNAs are translated into polypeptides in what may be considered the culmination of the primary process of gene action. The proteins so formed may act as enzymes, structural units, regulators of various metabolic processes by interacting with other proteins and genes, and essential agents in guiding and directing the processes of development.

The actual effects of gene action are recognized for the most part by noting the effects of gene mutation on the phenotype, but in complex multicellular organisms the final phenotypic effect observed superficially may be far removed from the initial action of the gene itself, for example, a change in shape of the ear or a change in eye color. Studies of a wide variety of different genetic strains of organisms ranging over phages and bacteria, plants and animals, including humans, have shown that the mutations of some genes are reflected in the qualitative alteration of the proteins they code for. These kinds of genes are called structural genes. Frequently the protein changes resulting from their mutation are simple substitutions of a single amino acid in the chain by another. The cause for this substitution can be traced back to a change in the genetic code. If a codon in the DNA of a gene coding for a specific polypeptide is altered by a base, for example, AAA → CAA, this mutation will result in the substitution of valine for the phenylalanine originally present at the specific site in the polypeptide chain coded for by this codon. A protein so changed, even though it may involve only one amino acid residue out of more than 100, may have no noticeable effects on the phenotype, or it may have drastic effects.

If there is no noticeable effect of mutation on a protein's action in forming the phenotype, the mutant allele may be called a neutral allele. However, even a single amino acid substitution may cause a protein to be completely inactive and, if the protein is an enzyme, create a genetic metabolic block. Over 100 different inherited blocks are known in humans, including phenylketonuria. The mutant protein may also be active, but its activity is altered so that it is less active than the wild type, or more active, or only active at certain temperatures or pH, or may be inhibited by substances not inhibitory to the wild-type enzyme. The range of possible effects is considerable, and by no means are all recognized and cataloged.

Some genes code for RNA that is not translated. The obvious ones are those that code for tRNA and rRNA, but other RNAs are also transcribed that do not appear to be transcribed and may have a role in the regulation of the activity of other genes. Finally, some genes code for proteins that act as regulators of the processes of metabolism and development. As yet, little is known about this class of genes in the eukaryotes, but they have been studied intensively in bacteria and phages. Mutations of genes coding for regulators may have profound effects on the course of development and, if they do not cause early lethality, they may result in the birth of a malformed individual.

Under certain conditions, an increase in the number of times a particular gene is present has a direct quantitative effect on one or more aspects of the phenotype. This effect is considered to be a manifestation of quantitative gene activity or gene dosage. It means that the gene does something, and that the higher the dose with which it is present, the greater the physiological or chemical end result. The best examples of this are found in the genetics of polyploid plants. In these plants, it is possible to increase the dosage of a particular allele from zero to four or more, when this is done, for example, with certain genes which control the production of the flower pigments of Dahlia, there is a demonstrable increase in the amount of pigment in the petals.

An increase in the dose of an active allele of a gene does not always cause an increase in the manifestation of a particular aspect of the phenotype. It may have quite the opposite effect, and cause a decrease. Plants and animals are also subject to aneuploidy, in which not all of the chromosomes of the genome are increased in number proportionately. Instead, only one of the chromosomes of the diploid set may be increased or decreased in number. In humans these conditions usually lead to the early death of the embryo. If the fetus reaches term and is born, it is always abnormal. A relatively common occurrence in humans is trisomy-21, which leads to Down syndrome.

Some heterozygous combinations of mutant alleles do not produce the phenotype expected from the phenotypes of the homozygotes. This is defined as a manifestation of allelic interaction. It is in contrast to those situations in which one allele is dominant over the other, so that the phenotype of the heterozygote is very similar or identical to that of the homozygous dominant. Also, it is different from those situations in which the two alleles show an additive effect, and the phenotype of the heterozygote is intermediate between those expected from the homozygotes. Diploid organisms heterozygous for two mutant alleles will produce two polypeptide species, one for each allele. The two proteins may interact and form hybrid multimers, which may be more or less active than the homomultimers formed by the polypeptides of each of the two genes alone.

The final phenotype of an organism is the resultant of the action of all the active genes in its cell or cells. These genes may act independently in producing their respective primary products, but the primary products, and the products of their activity, that is, enzymes and other macromolecules, interact at the level of extragenic metabolism to give the final phenotype. Thus, there is really no one gene determining the shape of an organ, or even the production of a certain pigment. These end products are determined by many genes acting together through their respective immediate and then succeeding interrelated products. The manifestation of these interactions, as determined initially by the results from breeding experiments and in a few cases from biochemical analysis, is called gene interaction. This term does not necessarily imply that the genes themselves interact. In general, the term is applied to apparent interactions between genes. Examples include: complementary genes, in which nonallelic genes are so directly involved in the formation of the same end product, or phenotype, that the mutation of any one of them to an inactive state will result in no end product or type effect; epistasis, in which one gene masks the effects of other genes that may be present; and suppressor genes, which cause a wild-type or normal phenotype despite the presence of nonallelic mutant genes.

The environment must be considered in any analysis of gene action, if it is desired to arrive at an understanding of how genes act toward the production of the phenotype. Practically, this is best done by keeping the environment as constant as possible while making studies of gene action. However, much also can be learned by varying the environmental conditions and keeping the genotype as constant as possible.

The environment of genes is a complex one. For convenience, two areas can be defined that immediately around the genes, the intracellular environment of the rest of the cell; and the extracellular and extraorganismal environment. The intracellular environment can be changed by the mutation of other genes, which may then modify the action of a gene under study. Gene interaction is thus seen to be in part an aspect of the study of the internal environment of the cell. Extracellular environmental factors, such as light and heat, may also influence the action of genes greatly.

Changes in phenotype which occur against a constant genetic background are in reality responses to the environment by the extragenic part of the living system. The genes themselves are not changed, as can be readily demonstrated by changing the environment back to the original condition, or by breeding the individual and showing that the offspring inherit the original parental genotype.

Genome

A genome is the complete set of genetic information in an organism. It provides all of the information the organism requires to function. In living organisms, the genome is stored in long molecules of DNA called chromosomes. Small sections of DNA, called genes, code for the RNA and protein molecules required by the organism. In eukaryotes, each cell's genome is contained within a membrane-bound structure called the nucleus. Prokaryotes, which contain no inner membranes, store their genome in a region of the cytoplasm called the nucleoid. The full range of RNA molecules expressed by a genome is known as its transcriptome, and the full assortment of proteins produced by the genome is called its proteome.

There are 23 pairs of chromosomes in the human genome. Between 1990 and 2003, all twenty-three pairs were fully sequenced through an international research undertaking known as the Human Genome Project. The study and analysis of genomes is called genomics.

DNA

DNA is the molecule that is the hereditary material in all living cells. Genes are made of DNA, and so is the genome itself. A gene consists of enough DNA to code for one protein, and a genome is simply the sum total of an organism's DNA.

DNA is long and skinny, capable of contorting like a circus performer when it winds into chromosomes. It's skinny as a whip and smart as one too, containing all the information necessary to build a living organism. In a very real sense, DNA is information.

DNA is a very large molecule, made up of smaller units called nucleotides that are strung together in a row, making a DNA molecule thousands of times longer than it is wide. Each nucleotide has three parts: a sugar molecule, a phosphate molecule, and a structure called a nitrogenous base. The nitrogenous base is the part of the nucleotide that carries genetic information, so the words "nucleotide" and "base" are often used interchangeably. The bases found in DNA come in four varieties: adenine, cytosine, guanine, and thymine—often abbreviated as A, C, G, and T, the letters of the genetic alphabet.

DNA is the Hereditary Material

DNA was largely ignored for decades after a German chemist, Friedrich Miescher, first isolated the white, slightly acidic substance from the nucleus of cells in 1869. No one knew what DNA's function was—in fact, some doubted that it had a function at all—so they pretty much left the stuff alone.

Very few people thought that DNA could be the hereditary material. Early studies of DNA suggested, erroneously, that the molecule was made up of the same sequence of four bases repeated over and over—ACGTACGTACGT for example. No one could imagine how such a monotonously simple molecule could contain the information necessary to build a living organism.

But during the 1930s and 1940s, new experiments began to suggest that DNA might, in fact, be important. It turned out that different strains of bacteria can exchange DNA and that when they do certain traits, such as the ability to cause disease in humans, can be passed from one strain of bacteria to another. Scientists also learned that when a virus infects a cell it injects its DNA into the cell, which then produces many copies of the virus, suggesting that DNA contains instructions for building viruses. And they found that different species of organisms have different proportions of bases in their DNA—one species might have DNA that is 30 percent A, 20 percent C, 20 percent G, and 30 percent T, while another might have 20 percent A, 30 percent C, 30 percent G, and 20 percent T. People began to think that genetic information might be written in the differences between the DNA bases of different species.

Structure of DNA

A DNA molecule is a double helix, a structure that looks much like a ladder twisted into a spiral. The sides of the ladder are made of alternating sugar and phosphate molecules, the sugar of one nucleotide linked to the phosphate of the next. DNA is often said to have a sugar and phosphate "backbone".

Each rung of the ladder is made of two nitrogenous bases linked together in the middle. The length of a DNA molecule is often measured in "base pairs," or bp—that is, the number of rungs in the ladder. Sometimes, this unit of measurement is shortened simply to "bases".

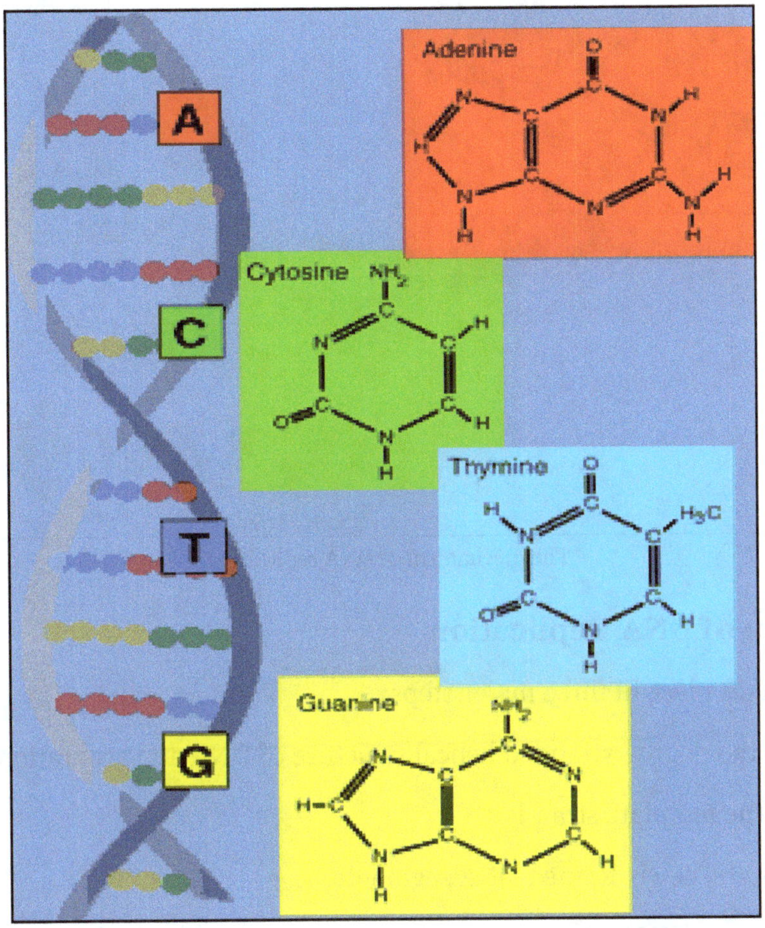

The structure of DNA was worked out by James D. Watson and Francis Crick. By the time they began their work in the early 1950s, it was clear that DNA is the hereditary material, and scientists were racing to find out more about the long-ignored molecule, picking apart the implications of each new detail. Everyone knew they couldn't really understand how DNA works until they understood how its nucleotide building blocks are put together.

DNA Replication

DNA replication is a semiconservative process where a parental strand (template) is used to synthesize a new complementary daughter strand using several protein elements which include enzymes and RNA molecules. DNA replication process uses DNA polymerase as the main enzyme for catalyzing the joining of deoxyribonucleoside 5′-triphosphates (dNTPs) forming a growing chain of DNA. Other proteins are also involved for initiation of the process and copying of DNA, along with proofreading capabilities to ensure the replication process takes place accurately. Therefore DNA replication is a process that produces identical helices of DNA from a single strand of the DNA molecule. DNA replication is an essential mechanism in enhancing cell growth, repair, and reproduction of an organism.

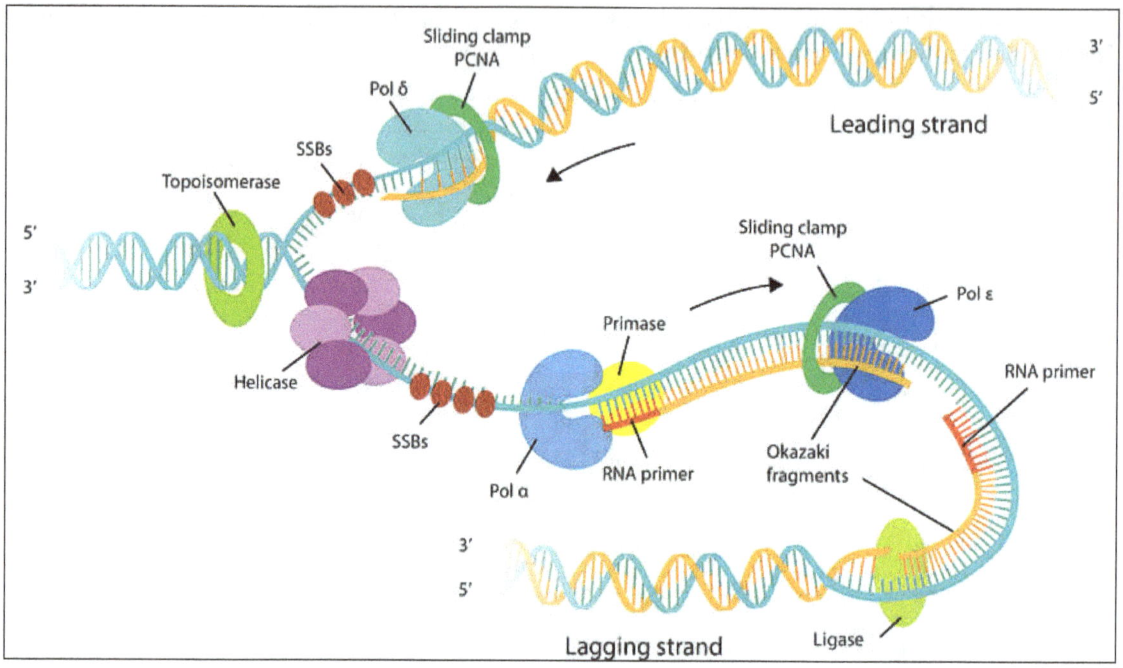

The mechanism of DNA replication.

The Mechanism of DNA Replication

DNA replication takes place in three major steps:

- Opening of the double-stranded helical structure of DNA and separation of the strands.
- Priming of the template strands.
- Assembly of the newly formed DNA segments.

During the separation of DNA, the two strands uncoil at a specific site known as the origin. With the involvement of several enzymes and proteins, they prepare (prime) the strands for duplication. At the end of the process, DNA polymerase enzyme starts to organize the assembly of the new DNA strands. These are the general steps of DNA replication for all cells but they may vary specifically, depending on the organism and cell type. Enzymes play a major role in DNA replication because they catalyze several important stages of the entire process. DNA replication is one of the most essential mechanisms of a cell's function and therefore intensive research has been done to understand its processes. The mechanism of DNA replication is well understood in Escherichia coli, which is also similar to that in eukaryotic cells. In E.coli, DNA replication is initiated at the oriClocus (oriC), to which DnaA protein binds while hydrolyzing of ATP takes place.

DNA Replication Enzymes and Proteins

- DNA Polymerase: DNA polymerases are enzymes used for the synthesis of DNA by adding nucleotide one by one to the growing DNA chain. The enzyme incorporates complementary amino acids to the template strand. DNA polymerase is found in both prokaryotic and eukaryotic cells. They both contain several different DNA polymerases responsible for different functions in DNA replication and DNA repair mechanisms.

- DNA Helicase Enzyme: This is the enzyme that is involved in unwinding the double-helical structure of DNA allowing DNA replication to commence. It uses energy that is released during ATP hydrolysis, to break the hydrogen bond between the DNA bases and separate the strands. This forms two replication forks on each separated strand opening up in opposite directions. At each replication fork, the parental DNA strand must unwind exposing new sections of single-stranded templates. The helicase enzyme accurately unwinds the strands while maintaining the topography on the DNA molecule.
- DNA Primase Enzyme: This is a type of RNA polymerase enzyme that is used to synthesize or generate RNA primers, which are short RNA molecules that act as templates for the initiation of DNA replication.
- DNA Ligase Enzyme: This is the enzyme that joins DNA fragments together by forming phosphodiester bonds between nucleotides.
- Exonuclease: These are a group of enzymes that remove nucleotide bases from the end of a DNA chain.
- Topoisomerase: This is the enzyme that solves the problem of the topological stress caused during unwinding. They cut one or both strands of the DNA allowing the strand to move around each other to release tension before it rejoins the ends. And therefore, the enzyme catalysts the reversible breakage it causes by joining the broken strands. Topoisomerase is also known as DNA gyrase in E. coli.
- Telomerase: This is an enzyme found in eukaryotic cells that adds a specific sequence of DNA to the telomeres of chromosomes after they divide, stabilizing the chromosomes over time.

DNA Replication Steps/Stages

Initiation

This is the stage where DNA replication is initiated. DNA synthesis is initiated within the template strand at a specific coding region site known as origins. The origin sites are targeted by the initiator proteins, which recruit additional proteins that help in the replication process to form a replication complex around the DNA origin. There are several origin sites on which DNA replication is initiated and they are all known as replication forks. The formed replication complex contains the DNA helicase enzyme whose function is to unwind the double helix, exposing the two strands, which act as templates for replication. The mechanism of DNA helicase enzyme is by hydrolyzing the ATP that is used to form the bonds between the nucleobases, thus breaking the bond that holds the two strands. Additionally, during initiation DNA primase enzyme synthesizes small RNA primers that kick-start the function of DNA polymerase. DNA polymerase enzyme functions by growing the new DNA daughter strand.

Elongation

This is the phase where the DNA polymerase grows the new DNA daughter strand by attaching to the original unzipped template strand and the initiating short RNA primer. The DNA polymerase is able to synthesize a new strand that matches the template, by extending the primer via the addition of free nucleotides to the 3′ end. One of the templates reads in the 3′ to 5′ direction, and therefore, the DNA polymerase synthesizes the new strand in the 5′ to 3′ direction, which is known as the leading strand. Along the template strand, DNA primase synthesizes a short RNA primer at the beginning of the template in the 5′ to 3′ direction, which initiates the DNA polymerase to continue synthesizing new nucleotides, extending the new DNA strand. The other template (5′ to 3′) is elongated in an antiparallel direction, by the addition of short RNA primers which are filled with other joining fragments, forming the newly formed lagging strand. These short fragments are known as the Okazaki fragments. The synthesis of the lagging strand is discontinuous since the newly formed strand is disjointed. The RNA nucleotides from the short RNA primers must be removed and replaced by DNA nucleotides, which are then joined by the DNA ligase enzyme.

Termination

After the synthesis and extension of both the continuous and discontinued stands, an enzyme knows as exonuclease removes all RNA primers from the original strands. The primers are replaced with the right nucleotide bases. While removing the primers, another type of exonuclease proofread the new stands, checking, removing, and replacing any errors formed during synthesis. DNA ligase enzyme joins the Okazaki fragments to form a single unified strand. The ends of the parent strand consist of a repetition of DNA sequences known as telomeres which act as protective caps at the ends of chromosomes preventing the fusion of nearby chromosomes. The telomeres are synthesized by a special type of DNA polymerase enzyme known as telomerase. It catalyzes the telomere sequences at the end of the DNA. On completion, the parent and complementary strand coil into a double helical shape, producing two DNA molecules each passing one strand from the parent molecule and one new strand.

Okazaki Fragments

The two DNA strands run in opposite or antiparallel directions, and therefore to continuously synthesize the two new strands at the replication fork requires that one strand is synthesized in the 5' to 3′ direction while the other is synthesized in the opposite direction, 3' to 5'. However, DNA polymerase can only catalyze the polymerization of the dNTPs only in the 5' to 3'direction. This means that the other opposite new strand is synthesized differently. But how? By the joining of discontinuous small pieces of DNA that are synthesized backward from the direction of movements of the replication fork. These small pieces or fragments of the new DNA strand are known as the Okasaki Fragments. The Okasaki fragments are then joined by the action of DNA ligase, which forms an intact new DNA strand known as the lagging strand. The lagging phase is not synthesized by the primer that initiates the synthesis of the leading strand. Instead, a short fragment of RNA serves as a primer (RNA primer) for the initiation of replication of the lagging strand. RNA primers are formed during the synthesis of RNA which is initiated de novo, and an enzyme known as primase synthesizes these short fragments of RNA, which are 3-10 nucleotides long and complementary to the lagging strand template at the replication fork. The Okazaki fragments are then synthesized by the extension of the RNA primers by DNA polymerase. However, the newly synthesized lagging strand is that it contains an RNA-DNA joint, defining the critical role of RNA in DNA replication.

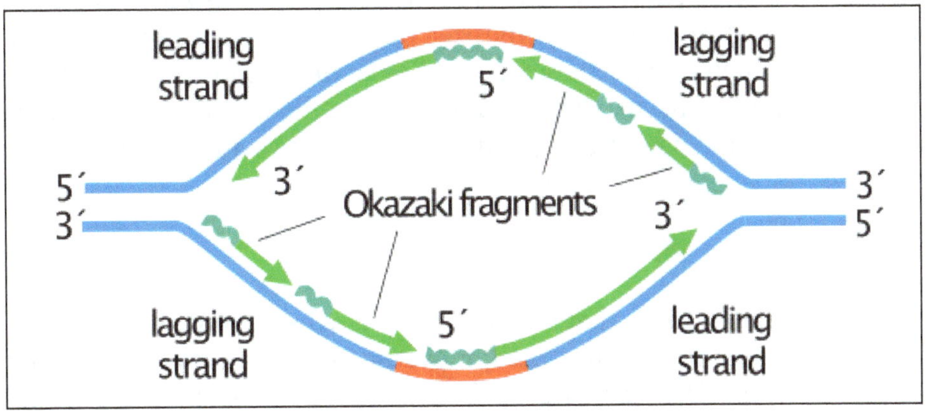

Replication Fork Formation and its Function

The replication fork is the site of active DNA synthesis, where the DNA helix unwinds and single strands of the DNA replicates. Several sites of origin represent the replication forks. The replication fork is formed during DNA strand unwinding by the helicase enzyme which exposes the origin of replication. A short RNA primer is synthesized by primase and elongation done by DNA polymerase. The replication fork moves in the direction of the new strand synthesis. The new DNA strands are synthesized in two orientations, i.e 3′ to 5′ direction which is the leading strand, and the 5′ to 3′ orientation which is the lagging strand. The two sides of the new DNA strand (leading and lagging strand) are replicated in two opposite directions from the replication fork. Therefore the replication fork is bi-directional.

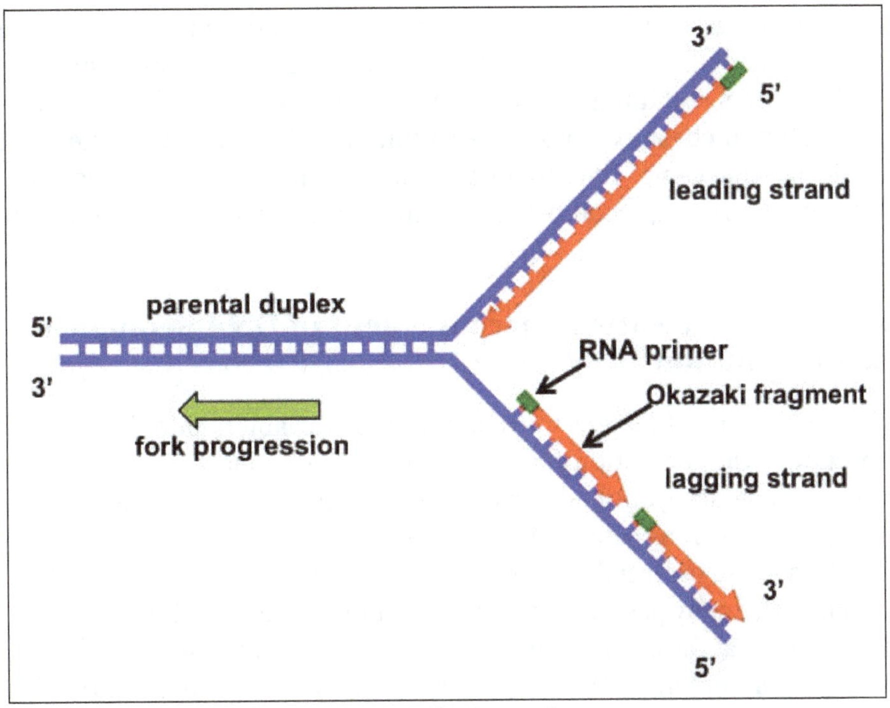

DNA Replication Fork.

Leading Strand

The leading strand is the new DNA strand that is continuously synthesized by the DNA polymerase enzyme. It is the simplest strand that is synthesized during replication. The synthesis starts after the DNA strand has unzipped and separated. This generates a short piece of RNA known as a primer, by the DNA primase enzyme. The primer binds to the 3′ end (start) of the strand, thus initiating the synthesize of the new strand (leading strand). The synthesis of the leading strand is a continuous process.

The Lagging Strand

This is the template strand (5′ to 3′) that is synthesized in a discontinuous manner by RNA primers. During the synthesis of the leading strand, it exposes small, short strands, or templates that are then used for the synthesis of the Okasaki fragments. The Okasaki fragments synthesize the lagging strand by the activity of DNA polymerase which adds the pieces of DNA (the Okasaki fragments) to the strand between the primers. The formation of the lagging strand is a discontinuous process because the newly formed strand (lagging strand) is the fragmentation of short DNA strands.

Recombination

Recombination is a process by which pieces of DNA are broken and recombined to produce new combinations of alleles. This recombination process creates genetic diversity at the level of genes that reflects differences in the DNA sequences of different organisms.

In eukaryotic cells, which are cells with a nucleus and organelles, recombination typically occurs during meiosis. Meiosis is a form of cell division that produces gametes, or egg and sperm cells. During the first phase of meiosis, the homologous pairs of maternal and paternal chromosomes align. During the alignment, the arms of the chromosomes can overlap and temporarily fuse, causing a crossover. Crossovers result in recombination and the exchange of genetic material between the maternal and paternal chromosomes. As a result, offspring can have different combinations of genes than their parents. Genes that are located farther apart on the same chromosome have a greater likelihood of undergoing recombination, which means they have a greater recombination frequency.

Genetic Recombination is the Rearrangement of DNA Sequences to Create New Genetic Information

Genetic recombination is one of the few ways to introduce variability into the genome. Recombination is most important in the generation of diversity among individuals in a population, although it is also a mechanism for generating antibody diversity in vertebrates. Recombination is also important in pathological processes such as cancer. Many viruses, including the cancer-causing retroviruses, rely on recombination for an essential part of their life cycle: the joining of the viral genome with the host genome. In addition, many parasites rely on recombination to alter the genes that code for components of their surface coats so as to escape the host's immune defenses. A final area of importance for genetic recombination centers on the use of recombination by researchers to artificially introduce gene segments into the genome of cells or whole organisms.

General Homologous Recombination is the Simplest and Most Studied Category of Recombination

General homologous recombination is the exchange of DNA strands that have a certain degree of complementarity. In homologous recombination the two DNA segments need not be completely identical, but only homologous, to recombine. Studies of homologous recombination in bacteria have identified a key protein, RecA, that catalyzes the exchange of genetic information. Many of the essential features of homologous recombination have been reconstituted with purified RecA protein. These biochemical studies have brought researchers to the point of asking two fundamental questions. First, how does RecA locate the homologous segments of DNA and bring them together to initiate the exchange? Second, how does RecA catalyze the extension of the initial point of DNA exchange so that hundreds of thousands of nucleotides of DNA are recombined? We know that this process requires hydrolysis of the high-energy compound adenosine triphosphate.

In eukaryotic systems, the stage is set for major research advances. Molecular genetic techniques should expedite the identification, isolation, and characterization of the genes that encode the recombination apparatus. Because of the ease of genetic manipulation, research on the lower eukaryotes may lead the way in this effort. One hopes, however, that conservation of gene structure will permit identification of related genes in higher eukaryotes. The biochemical study of meiotic components—the complex that aligns homologous chromosomes, DNA strand transferases that initiate exchange, and DNA resolvases that terminate exchange—is in its infancy, but recent progress in yeast should attract more workers to the field.

In addition to understanding the basic mechanism, we need to learn more about important control features of meiotic recombination. For example, failure of homologues to separate at meiosis is the basis of trisomy diseases like Down's syndrome. A second item for further study is the means by which cells prevent recombination between repeated sequences. Such repetitive DNA is a hallmark of higher eukaryotic cells and presents a potentially devastating opportunity for a homologous recombination system. If left unchecked, recombination between repeated sequences could scramble the genome. We need to know how this is limited and whether breakdowns in this control are responsible for some human disease.

A separate issue in the study of general recombination of eukaryotes concerns somatic cells. When recombination has been studied in living organisms (usually Drosophila and yeast), somatic recombination is less frequent than recombination in germline cells and seems to be primarily involved in DNA repair. Nevertheless, many scientists have demonstrated the efficient recombination of homologous segments of DNA when they are artificially introduced into somatic cells in culture. Although ongoing studies attempt to define the mechanism of this recombination, the most exciting possibility is its potential use to manipulate the genetic content of cultured cells. Research will focus on controlling the frequency and fidelity of recombination so that investigators can introduce genes into cultured cells and replace existing genes with modified ones. This sequence of manipulations is readily accomplished in bacteria, yeast, and slime molds, where the technique has been of tremendous value in the genetic analysis of complex loci. Formidable problems arise in extending this technology to higher organisms became of the complexity of their genomes. However, several initial successes in targeting have opened the way to a thorough attack on the problem.

Site-specific Recombination has been Studied for the Past 20 Years and Many Systems have been Studied in Depth

In the 1960s it became clear that efficient recombination could take place between segments of DNA that lacked sufficient homology to undergo general homologous recombination. Two features are required for such nonhomologous recombination—the presence of special DNA sites in at least one partner in the recombination process and the activity of specific enzymes (recombinases) that act at these sites. The rearrangements of the genome promoted by site-specific recombination are mechanically fascinating. One wants to know how small segments of the genome are identified, brought close together, and rearranged with high efficiency and fidelity. Moreover, site-specific recombination is a critical step in many important biological processes. As a result, the study of site-specific recombination has emerged as a major area of concentration of biological research that will continue to grow rapidly in the next 5 to 10 years.

The molecular mechanism of these reactions has been worked out by using the molecules and mutant strains identified by genetic analysis to reconstitute the recombination event in a cell-free (in vitro) system. Site-specific recombination can be divided into broad mechanistic categories according to the degree to which both partners must have specific sites and on the amount of DNA synthesis that accompanies the reaction. In some systems, recombination uses a completely conservative mechanism in which the participating DNAs are simply broken, rearranged, and rejoined. In other systems, complete replicas of the DNA must be synthesized during the recombination event. The integration of the bacterial virus called lambda into the E. coli genome is a reaction with high target specificity and strict conservation of the parental DNA. This complete reaction has been reconstituted in vitro. By combining topological investigations with biochemical and molecular genetic studies, the field is reaching the goal of elaborating plausible models for complex processes such as synapsis, the juxtaposition of DNA segments that is a prerequisite for efficient, precise recombination. In the future, structural studies using x-ray diffraction and nuclear magnetic resonance techniques should be able to add atomic details to the understanding of these processes that is being deduced from molecular biological studies.

In contrast to the conservative reactions (no DNA replication), study of the mechanism of the replicative site-specific recombination (which requires DNA replication) is in its infancy. The transposition of the genome of the bacterial virus mu, is the only example that has been reconstituted in vitro. Several features of the DNA sequence of donors and targets suggest close parallels between mu transposition on the one hand and other rearrangements such as the integration of retroviruses (RNA viruses that replicate by way of a DNA intermediate that is inserted into the host's chromosome) and the recombination of antibody genes. The pioneering work on mu transposition has thus opened the way for rapid progress in the biochemistry of these eukaryotic recombination events.

Another important research area, control of the recombination process, has already been extensively studied in prokaryotic and lower eukaryotic systems. Here, the key event seems to be control of the synthesis of a specific recombinase. In some systems, such as the integration of lambda and sexual differentiation of yeast, such control is elicited through a complex chain of events that couples the synthesis of the recombinase to related events in the life cycle of organisms. In higher eukaryotes, we know almost nothing about the control of recombination. We need to learn whether these events are regulated and, if so, by what mechanism. For the immune system, some experiments suggest that accessibility of the recombining sites rather than the availability of the recombinase may be critical. The finding that several human tumors result from an inappropriate rearrangement of immune elements highlights the importance of understanding the control of site-specific recombination in higher organisms.

References

- Genes, biology: byjus.com, Retrieved 03, August 2020
- Recombination-is-a-process-by-DNA-sequences-of-different-organisms, recombination, scitable: nature.com, Retrieved 12, June 2020
- Dna-replication: microbenotes.com, Retrieved 05, July 2020
- The-primary-action-of-a-primary-process-of-gene-action: encyclopedia2.thefreedictionary.com, Retrieved 25, July 2020
- Mitochondria, cells: micro.magnet.fsu.edu, Retrieved 27, May 2020

Chapter 3

Human Biology

Human biology deals with human beings and population. All aspects of human functionality, be it genetics, anatomy, physiology, anthropology, etc. are studied under human biology. It is an interdisciplinary study with close ties to biomedical sciences, biomedical anthropology and many other studies related to the human body. This chapter provides a detailed explanation of the mentioned fields.

Human Body

Human biology is the branch of biology that focuses on human beings and human populations; it encompasses all aspects of the human organism including genetics, ecology, anatomy and physiology, anthropology, and nutrition, among others. Human biology is related to other fields of biology such as medicine, primate biology, and biological anthropology. Human body is the physical substance of the human organism, composed of living cells and extracellular materials and organized into tissues, organs, and systems.

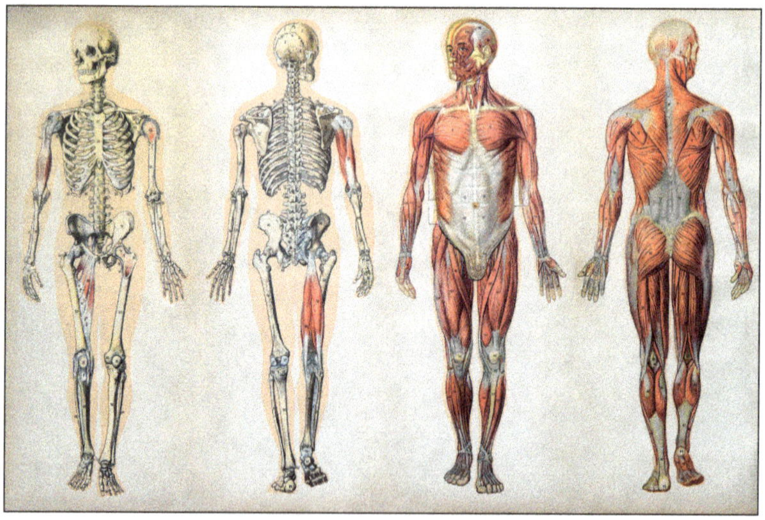

Human body: Vintage anatomy charts of the human body showing the skeletal and muscle systems.

Humans are, of course, animals—more particularly, members of the order Primates in the subphylum Vertebrata of the phylum Chordata. Like all chordates, the human animal has a bilaterally symmetrical body that is characterized at some point during its development by a dorsal supporting rod (the notochord), gill slits in the region of the pharynx, and a hollow dorsal nerve cord. Of these features, the first two are present only during the embryonic stage in the human; the notochord is replaced by the vertebral column, and the pharyngeal gill slits are lost completely. The dorsal nerve cord is the spinal cord in humans; it remains throughout life.

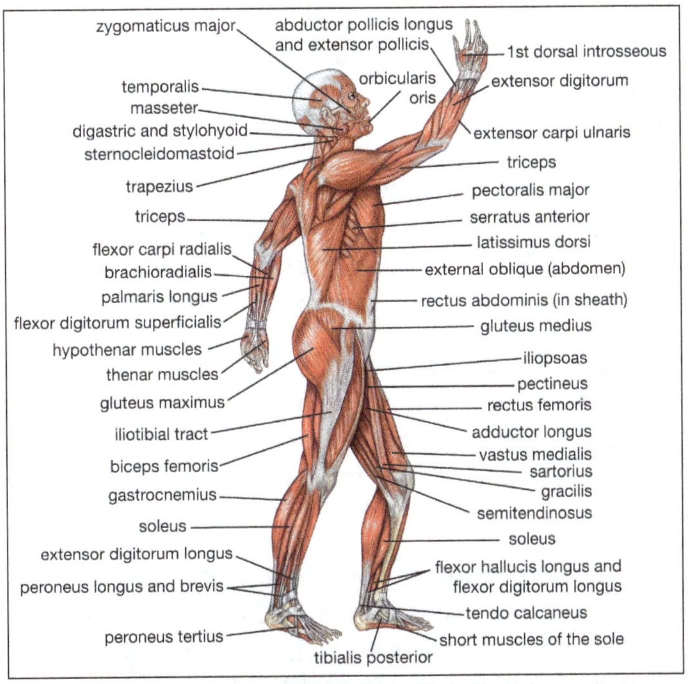

Human muscular system: Lateral view of the human muscular system.

Characteristic of the vertebrate form, the human body has an internal skeleton that includes a backbone of vertebrae. Typical of mammalian structure, the human body shows such characteristics as hair, mammary glands, and highly developed sense organs.

Beyond these similarities, however, lie some profound differences. Among the mammals, only humans have a predominantly two-legged (bipedal) posture, a fact that has greatly modified the general mammalian body plan. (Even the kangaroo, which hops on two legs when moving rapidly, walks on four legs and uses its tail as a "third leg" when standing.) Moreover, the human brain, particularly the neocortex, is far and away the most highly developed in the animal kingdom. As intelligent as are many other mammals—such as chimpanzees and dolphins—none have achieved the intellectual status of the human species.

Chemical Composition of the Body

Chemically, the human body consists mainly of water and of organic compounds—i.e., lipids, proteins, carbohydrates, and nucleic acids. Water is found in the extracellular fluids of the body (the blood plasma, the lymph, and the interstitial fluid) and within the cells themselves. It serves as a solvent without which the chemistry of life could not take place. The human body is about 60 percent water by weight.

Lipids—chiefly fats, phospholipids, and steroids—are major structural components of the human body. Fats provide an energy reserve for the body, and fat pads also serve as insulation and shock absorbers. Phospholipids and the steroid compound cholesterol are major components of the membrane that surrounds each cell.

Proteins also serve as a major structural component of the body. Like lipids, proteins are an important constituent of the cell membrane. In addition, such extracellular materials as hair and nails are composed of protein. So also is collagen, the fibrous, elastic material that makes up much of the body's skin, bones, tendons, and ligaments. Proteins also perform numerous functional roles in the body. Particularly important are cellular proteins called enzymes, which catalyze the chemical reactions necessary for life.

Carbohydrates are present in the human body largely as fuels, either as simple sugars circulating through the bloodstream or as glycogen, a storage compound found in the liver and the muscles. Small amounts of carbohydrates also occur in cell membranes, but, in contrast to plants and many invertebrate animals, humans have little structural carbohydrate in their bodies.

Nucleic acids make up the genetic materials of the body. Deoxyribonucleic acid (DNA) carries the body's hereditary master code, the instructions according to which each cell operates. It is DNA, passed from parents to offspring, that dictates the inherited characteristics of each individual human. Ribonucleic acid (RNA), of which there are several types, helps carry out the instructions encoded in the DNA.

Along with water and organic compounds, the body's constituents include various inorganic minerals. Chief among these are calcium, phosphorus, sodium, magnesium, and iron. Calcium and phosphorus, combined as calcium-phosphate crystals, form a large part of the body's bones. Calcium is also present as ions in the blood and interstitial fluid, as is sodium. Ions of phosphorus, potassium, and magnesium, on the other hand, are abundant within the intercellular fluid. All of these ions play vital roles in the body's metabolic processes. Iron is present mainly as part of hemoglobin, the oxygen-carrying pigment of the red blood cells. Other mineral constituents of the body, found in minute but necessary concentrations, include cobalt, copper, iodine, manganese, and zinc.

Organization of the Body

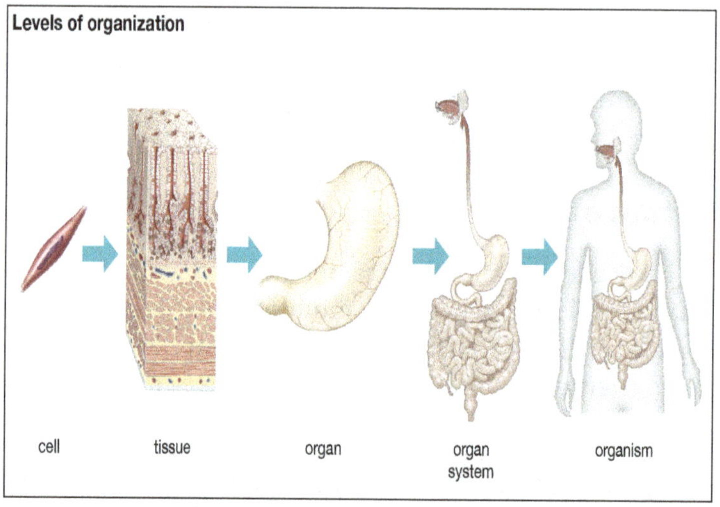

Multicellular organism: Organization.

The cell is the basic living unit of the human body—indeed, of all organisms. The human body consists of trillions of cells, each capable of growth, metabolism, response to stimuli, and, with some exceptions, reproduction. Although there are some 200 different types of cells in the body, these can be grouped into four basic classes. These four basic cell types, together with their extracellular materials, form the fundamental tissues of the human body: epithelial tissues, which cover the body's surface and line the internal organs, body cavities, and passageways; muscle tissues, which are capable of contraction and form the body's musculature; nerve tissues, which conduct electrical impulses and make up the nervous system; and connective tissues, which are composed of widely spaced cells and large amounts of intercellular matrix and which bind together various body structures. (Bone and blood are considered specialized connective tissues, in which the intercellular matrix is, respectively, hard and liquid).

The diagram shows five levels of organization in a multicellular organism. The most basic unit is the cell; groups of similar cells form tissues; groups of different tissues make up organs; groups of organs form organ systems; cells, tissues, organs, and organ systems combine to form a multicellular organism. The next level of organization in the body is that of the organ. An organ is a group of tissues that constitutes a distinct structural and functional unit. Thus, the heart is an organ composed of all four tissues, whose function is to pump blood throughout the body. Of course, the heart does not function in isolation; it is part of a system composed of blood and blood vessels as well. The highest level of body organization, then, is that of the organ system.

The body includes nine major organ systems, each composed of various organs and tissues that work together as a functional unit. The integumentary system, composed of the skin and associated structures, protects the body from invasion by harmful microorganisms and chemicals; it also prevents water loss from the body. The musculoskeletal system (also referred to separately as the muscle system and the skeletal system), composed of the skeletal muscles and bones (with about 206 of the latter in adults), moves the body and protectively houses its internal organs. The respiratory system, composed of the breathing passages, lungs, and muscles of respiration, obtains from the air the oxygen necessary for cellular metabolism; it also returns to the air the carbon dioxide that forms as a waste product of such metabolism.

The circulatory system, composed of the heart, blood, and blood vessels, circulates a transport fluid throughout the body, providing the cells with a steady supply of oxygen and nutrients and carrying away waste products such as carbon dioxide and toxic nitrogen compounds. The digestive system, composed of the mouth, esophagus, stomach, and intestines, breaks down food into usable substances (nutrients), which are then absorbed from the blood or lymph; this system also eliminates the unusable or excess portion of the food as fecal matter. The excretory system, composed of the kidneys, ureters, urinary bladder, and urethra, removes toxic nitrogen compounds and other wastes from the blood. The nervous system, composed of the sensory organs, brain, spinal cord, and nerves, transmits, integrates, and analyzes sensory information and carries impulses to effect the appropriate muscular or glandular responses. The endocrine system, composed of the hormone-secreting glands and tissues, provides a chemical communications network for coordinating various body processes. The reproductive system, composed of the male or female sex organs, enables reproduction and thereby ensures the continuation of the species.

Basic Form and Development

In general structure, the human body follows a plan that can be described as a cylinder enclosing two tubes and a rod. This body plan is most clearly evident in the embryo; by birth, the plan is apparent only in the trunk region—i.e., in the thorax and abdomen.

The body wall forms the cylinder. The two tubes are the ventrally located alimentary canal (i.e., the digestive tract) and the dorsally located neural tube (i.e., the spinal cord). Between the tubes lies the rod—the notochord in the embryo, which becomes the vertebral column prior to birth. (The terms dorsal and ventral refer respectively to the back and the front, or belly, of an animal).

Within the embryo, the essential body parts are: the outer enclosing epidermal membrane (in the embryo called ectoderm); the dorsal neural tube; the supporting notochord; the ventral alimentary tube, which becomes the lining of the stomach and intestine (in the embryo called endoderm); the intermediate mass (in the embryo called mesoderm); and a rather fluid tissue that fills the interspaces, derived from the mesoderm and in the embryo called mesenchyme. Everything in the body derives from one of these six embryonic parts.

The mesoderm constitutes a considerable pad of tissue on each side of the embryo, extending all the way from the back to the front sides of the body wall. It is hollow, for a cleftlike space appears in it on each side. These are the right and left body cavities. In the dorsal part of the body they are temporary; in the ventral part they become permanent, forming the two pleural cavities, which house the lungs; the peritoneal cavity, which contains the abdominal organs; and the pericardial cavity, which encloses the heart. The dorsal part of the mesoderm becomes separated from the ventral mesoderm and divides itself into serial parts like a row of blocks, 31 on each side. These mesodermal segments grow in all directions toward the epidermal membrane. They form bones, muscles, and the deeper, leathery part of the skin. Dorsally they form bony arches protecting the spinal cord, and ventrally the ribs protecting the alimentary canal and heart. Thus they form the body wall and the limbs—much the weightier part of the body. They give the segmental character to the body wall in neck and trunk, and, following their lead, the spinal cord becomes correspondingly segmented. The ventral mesoderm is not so extensive; it remains near the alimentary tube and becomes the continuous muscle layer of the stomach and intestine. It also forms the lining of the body cavities, the smooth, shining, slippery pleura and peritoneum. The mesenchyme forms blood and lymph vessels, the heart, and the loose cells of connective tissues.

The neural tube itself is formed from the ectoderm at a very early stage. Anteriorly (i.e., toward the head) it extends above the open end of the cylinder and is enlarged to form the brain. It is not in immediate contact with the epidermis, for the dorsal mesoderm grows up around it and around the roots of the cranial nerves as a covering, separating the brain from the epidermis. Posteriorly the neural tube terminates in the adult opposite the first lumbar vertebra.

If the cylindrical body wall is followed headward, it is found to terminate ventrally as the tongue, dorsally in the skull around the brain, ears, and eyes. There is a considerable interval between eyes and tongue. This is occupied partly by a deep depression of the epidermis between them, which dips in to join the alimentary tube (lining of the mouth). Posteriorly the ventral body wall joins the dorsal at the tailbone (coccyx), thus terminating the body cavities.

Headward, the alimentary tube extends up in front of the notochord and projects above the upper part of the body wall (tongue) and in front of and below the brain to join the epidermal depression. From the epidermal depression are formed the teeth and most of the mouth lining; from the upper end of the alimentary canal are formed the pharynx, larynx, trachea, and lungs. The alimentary canal at its tail end splits longitudinally into two tubes—an anterior and a posterior. The anterior tube becomes the bladder, urethra, and, in the female, the lining of the vagina, where it joins a depression of the ectoderm. The posterior (dorsal) tube becomes the rectum and ends just in front of the coccyx by joining another ectodermal depression (the anus).

Effects of Aging

As the human body ages it undergoes various changes, which are experienced at different times and at varying rates among individuals. The skin is one of the most accurate registers of aging. It becomes thin and dry and loses elasticity. Patches of darker pigmentation appear, commonly called liver spots, though they have no relation to that organ. Hair grays and thins. Wounds take longer to heal; some reparations take five times as long at 60 as at 10 years of age. Sensory fibers in spinal nerves become fewer; the ganglion cells become pigmented and some of them die. In the auditory apparatus some nerve cells and fibers are lost, and the ability to hear high notes diminishes. In the eye the lens loses its elasticity.

Human aging: Most people start to feel the effects of aging after they are well into adulthood.

Organs such as the liver and kidneys lose mass with age and decline in efficiency. The brain is somewhat smaller after the age of 40 and shrinks markedly after age 75, especially in the frontal and occipital lobes. This shrinkage is not, however, correlated with declines in mental capacity. Intellectual declines in the elderly are the consequence of underlying disease conditions, such as Alzheimer disease or cerebrovascular disease.

The bones become lighter and more brittle because of a loss of calcium. This loss in bone mass is greater in women than men after the fifth decade. In joints the cartilage covering the ends of bone becomes thinner and sometimes disappears in spots, so bone meets bone directly and the old joints creak. Compression of the spinal column can lead to a loss of height. Muscular strength decreases but with marked individual variability.

The arteries become fibrous and sclerosed. Because of decreasing elasticity, they tend to become rigid tubes. Fatty spots, which appear in their lining even in youth, are always present in old age. In vitro experiments indicate that the body's cells are programmed to undergo a finite number of divisions, after which time they lose their reproductive capacity. Thus, the potential longevity of the human body—about 100 years—seems to be encoded within the very cells of the body.

Change Incident to Environmental Factors

Although the basic form of the human body was established in human anthropoid ancestors, evolutionary adaptations to different environments are apparent among various human populations.

For example, physical adaptations in humans are seen in response to extreme cold, humid heat, and high altitudes.

Extreme cold favors short, round persons with short arms and legs, flat faces with fat pads over the sinuses, narrow noses, and a heavier than average layer of body fat. These adaptations provide minimum surface area in relation to body mass for minimum heat loss, minimum heat loss in the extremities (which allows manual dexterity during exposure to cold and guards against frostbite), and protection of the lungs and base of the brain against cold air in the nasal passages.

In hot climates the problem is not in maintaining body heat but in dissipating it. Ordinarily the body rids itself of excess heat by sweating. In conditions of humid heat, however, the humidity of the surrounding air prevents the evaporation of perspiration to some extent, and overheating may result. Hence, the heat-adapted person in humid climates is characteristically tall and thin, so that there is maximum surface area for heat radiation. The person living in hot climates has little body fat; often a wide nose, since warming of the air in the nasal passages is not desirable; and, usually, dark skin, which provides a shield from harmful solar radiation.

High altitudes demand a degree of cold adaptation, as well as adaptation for low air pressure and the consequent low oxygen. This adaptation is accomplished by an increase in lung tissue generally. Despite the fact that the general shape and size of the body and its parts are determined by heredity, the body can undergo some modifications in response to present conditions. Thus, a person who moves from a home at sea level to one at mountain altitudes will experience an increase in the number of red blood cells; this increase helps compensate for the lower oxygen levels of the new environment. Similarly, a light-skinned individual who moves to a hot tropical region will develop increased pigmentation in the skin. In such situations, the resultant form is seldom perfect for the new conditions, but it is adapted to present needs well enough to maintain life with the least waste of energy.

Human Proteins

The human body is made up almost entirely of protein. Protein is the main component of muscles, bones, organs, skin, and nails. Excluding water, muscles are composed of about 80% protein, making this nutrient especially important for athletes. The body breaks down consumed protein into amino acids, and absorbs it. It is used to build muscles and organs, to make hormones and antibodies, to be stored as fat, and to be burned as energy.

Proteins are large, complex molecules that play many critical roles in the body. They do most of the work in cells and are required for the structure, function, and regulation of the body's tissues and organs. Proteins are made up of hundreds or thousands of smaller units called amino acids, which are attached to one another in long chains. There are 20 different types of amino acids that can be combined to make a protein. The sequence of amino acids determines each protein's unique 3-dimensional structure and its specific function.

Genome sequencing deciphered the number of protein-coding genes, establishing an initial estimation of complexity associated with human molecular biology. The next step is to obtain similar benchmarks at the proteome level. Nevertheless, considerable efforts are still required for exploring the space (or size) of the human proteome, as a compulsory constellation of molecular profiles of different tissues and organs. The human proteome is quite a dynamic entity and this property should be considered in two dimensions. The first is to estimate the number of different protein types (proteome width), as well as measure protein copies number in particular tissues (proteome depth).

Following the hypothesis of "one gene = one protein," there should be at least ~20,000 nonmodified (canonical) human proteins. Taking into account products of alternative splicing (AS), those containing single amino acid polymorphisms (SAPs) arising from nonsynonymous single-nucleotide polymorphisms (nsSNPs), and those that undergo PTMs, as many as 100 different proteins can potentially be produced from a single gene. Of the many different terms proposed to describe protein variants, here, we chose "protein species" or "proteoforms".

Experimental validation of protein species is limited by the analytical sensitivity of proteomic technology. This means that the sensitivity of the technology determines the ability to detect rare protein species. This limitation originates from the basic difference between genomics and proteomics. Genomics relies upon PCR to amplify DNA or RNA molecules in a biological sample to concentrations above the detection threshold. However, there currently exists no comparable high-throughput technology capable of multiplying the copies of a single protein.

The 100% coverage of protein sequence using bottom-up MS is not attainable; thus, it is impossible to detect all potential protein species expressed from the same gene. Generally, proteome investigations are focused on the master proteins resembling at least one of the many possible proteoforms, coded by the gene and containing at least one MS-detectable proteotypic peptide. The sequence could be modified or nonmodified, so this means that the master protein could be present as a single protein or as a set of proteins. Master proteome of a single chromosome is the result of the identification and measurement of all master proteins encoded by the chromosome and expressed in the selected type of biological material. For experimental validation of proteoforms, the targeted MS analysis should be performed in order to probe candidate sequence alteration. The bioinformatic analysis of the diversity of protein species was anticipated to create the backbone for the future experimental exploration of the proteome space.

How Many Different Proteins are Necessary to Support Human Function?

The number of different proteins comprising the human proteome is a core proteomics issue. Researchers propose numbers between 10,000 and several billion different protein species. Here, we describe the theoretical prediction for the number of different proteoforms that might arise from AS, SAP, or PTM events.

The data was derived from neXtProt, which contains only human proteins and their modifications and sequence features. The neXtProt annotation of AS, SAP, and PTM originated from biocuration of the data from repositories, literature, and prediction tools. Information on possible protein sequence variability is represented as the number of AS variants, nsSNPs/SAPs, and PTMs per gene.

Our assumption was that database extension and annotation are a constituent process, whose rate is mostly limited by the number of the researchers and annotators around the world. The rate is slightly dependent on the capacity of communication channel and information accessibility, as these were not changed too much for 10–15 years for the needs of PubMed or UniProt users.

Therefore, the extension of the number of annotations in a certain database would generally be affected by the technology achievements, gained by increasing the sensitivity/throughput of the bioanalytical method.

From the above, we proposed that the volume of representative data uploaded to UniProt each year from 2005 was sufficient to calculate the average number of protein variants per one gene and the numbers for each type of variation. Interestingly, since 2010, the average number of modifications per one gene has remained nearly the same, despite the continuous increase in reviewed annotations. The average number of modifications specifically by AS (40% reviewed annotations out of all data records), SAP (60% reviewed annotations), or PTM (37%) remains almost unchanged.

The saturation in the number of annotations for genome-dependent SAPs, transcription-dependent ASs, and posttranslational-dependent PTMs is quite remarkable. While PTM determination depends upon the sensitivity of protein analytics, SAP and AS detection have virtually no limitations in sensitivity and are actively accumulated via large-scale projects. Despite such differences, all of the technologies have synchronously acquired saturation levels, indicating balance between data derived from using standard protein-chemistry techniques (accumulated over the last 50 years) and data derived from high-throughput next-generation sequencing (NGS).

For estimating the potential number of proteins, three different cases of combination of PTM, SAP, and AS events were considered. Combinatorial variations were ignored, since there are no systematic experimental data describing the cooccurrence of various modification types in the protein species. This is just one of the possible ways for solving the problem of how to estimate a potential number of proteins based on the data of protein variance that has already been accumulated on the postgenomic knowledge bases. Equation Nps=N∗(ASav+SAPav+PTMav), assumes that PTMs appear exclusively in the canonical sequences of proteins, but not in splice variants. Equation Nps=(N+AS)∗(SAPav+PTMav), assumes that PTMs and SAPs can occur both in proteins encoded by canonical sequences and in splice variants. Equation Nps=N∗ASav∗SAPav∗PTMav, assumes that all modification types (AS, SAP, and PTM) occur independently. Hence,

$$Nps = N * (ASav + SAPav + PTMav),$$

$$Nps = (N + AS) * (SAPav + PTMav),$$

$$Nps = N * ASav * SAPav * PTMav,$$

where Nps represents the number of protein species, N represents the total number of protein encoding genes, AS is the number of species produced by alternative splicing, ASav is the average number of splice variants per one protein encoding genes, SAPav is the average number of nsSNPs, and PTMav is the average number of PTM events per one protein encoding gene.

Generally, SAPs are predetermined at the DNA level, and AS arises from modifications at the mRNA level, while PTMs occur at the protein level. These three processes cannot be viewed as independent events, given that there is an intrinsic relationship between the processes of gene expression, transcription, and translation, aimed at regulating and preserving a cell. Furthermore, enriching MS/MS searches through a database containing all possible combinations of protein variations would lead to combinatorial collapse, despite the type of approach used.

The neXtProt (ver. 2015_06) search for protein AS modifications revealed 21,921 AS variants in 10,519 protein-coding genes (2.1 ± 0.1 variants/gene, including one canonical sequence). The greatest number of modified forms (434,398, without cancer-related items derived from the COSMIC cancer mutation database) was due to the emergence of SAPs resulting from nsSNPs in 18,986 protein-coding genes (22.1 ± 3.9 variants/gene). PTMs added 6.6 ± 0.8 modified proteins/gene (94,036 PTMs in 14,006 protein-coding genes). Applying these numbers to the equations (N = 20,043), we estimate that in humans there exist 0.62 or 0.88 or 6.13 million protein species.

The above results were matched to the data on AS- and SAP-derived variances obtained from our NGS results of liver tissue transcriptome profiling. According to NGS results, the average number of detected splice variants was 1.3 per protein-coding gene (or 2.3 per gene including canonical variant), which is comparable to neXtProt data. The average number of SAP-containing proteoforms was ~1.4 per one gene, so much lower than that calculated from neXtProt data. These differences relate to the fact that neXtProt provides information from many different experiments ("aggregate human population"), while specific NGS data indicates SAP events for an individual sample or tissue (individual variances).

As proteomic knowledge bases consolidate information regarding protein variability in the human population, several million different proteins will ultimately populate the "aggregated" human proteome. To decipher variability inherent in predicting proteome space for an individual, more precise estimation of the numbers of AS- and SAP-contained proteins can be achieved using results of transcriptome profiling of specific tissue samples.

How Many Protein Species are Detectable Today?

According to the Plasma Proteome Database, 10.5 thousand blood-plasma proteins have been detected and less than 10% (1278 of 20,043 human proteins) have been measured in a quantitative manner. The primary issue concerning experimental validation of existing sets of theoretically predicted proteins is the limit of analytical sensitivity of proteomic technology. Analytical sensitivity is determined by instrument-dependent detection limit and biomaterial-dependent dynamic protein concentration ranges. Blood plasma is a complex mixture with a dynamic range of protein concentrations varying by >10 orders of magnitude, while the protein concentration range of tissue or cell lines is within seven orders of magnitude. The challenge is in detecting low- and ultralow-abundance species with concentrations $<10^{-12}$ M in the presence of high-copied protein molecules at concentrations $>10^{-6}$ M.

Assuming the ultrasensitive capacity of oligonucleotide analytics, it is instructive to consider that transcriptome research results are often determined based on copies of RNA molecules rather than concentrations. Operating at low- ($<10^{-12}$ M) and ultralow ($<10^{-15}$ M) concentrations of proteins implies that quantifying protein in copy numbers rather than in concentration units enables comparison of transcriptomic and proteomic results.

Proteins are commonly quantified in the proteomics field by the concentration in the biological sample, C, reported as mol/L (molarity, M). The corresponding number of protein copies, N, in 1 L can be calculated out of concentration units as follows:

$$N = \left(\frac{C * V}{R_A} \right),$$

$$C = \left(\frac{m}{M_w * V} \right),$$

where R_A represents the reverse Avogadro's number, 10^{-24} M, V represents the sample volume, m represents the protein content, and M_w represents molecular weight of the protein.

Formulas above address the major challenge of proteomics: shift from concept of the concentration units to counting single biomacromolecules in a sample (tissue). The triple-quadrupole mass spectrometer makes it possible to achieve 10^{-14} M sensitivity for targeted proteins. The sensitivity of SRM protein detection can be further increased up to 10^{-16} M by irreversible chemical binding proteins from large volumes of biological samples (it is not intended to state that all proteins measured were determined with such sensitivity; results of measurements can vary by several orders of magnitude due to different physicochemical properties of proteotypic peptides).

In the context of proteome width, the targeted approach is limited by the need to measure only proteoforms exhibiting a priori assumption of proteotypic peptides, which correctly resemble PTM, SAP, or AS events. In contrast to shotgun MS, SRM cannot discover new, unexpected protein species. Possibilities of top-down and bottom-up MS approaches to address the microheterogeneity of the human proteome. Targeted SRM is readily available for detecting SAPs in association with disease, including obesity/diabetes and cancer. For example, SRM/MRM method was applied to measure the quantities of splice forms: three isoforms for transforming growth factor were measured by SRM at concentration level of 10^{-11} M in mouse plasma and human saliva. Another example, osteopontin isoforms, was measured using the SRM assay and revealed that level of isoform was significantly higher for non-small cell lung carcinoma compared with the control group ($7*10^{-10}$ versus $30*10^{-10}$ M). The application of targeted MS for the detection of PTMs was illustrated for protein glycosylation: N-glycosides were detected in human plasma at a sensitivity level of 10^{-11} M and ubiquitination. From these pilot studies, it follows that the vast majority of predictable proteoforms seem to be present in the concentrations below limit of detection. Further increase in sensitivity of analytical methods is important to uncover diagnostically relevant proteoforms in human biosamples.

Since it was shown that the set of proteins encoded by any human chromosome constitutes a representative portion for the whole human proteome, high-, medium-, and low-copied protein species can be evaluated by sampling master proteins encoded by a single chromosome. As an example of a chromosome-centric proteomic map, we uploaded data from PASSEL (PASSEL IDs: PASS00278, PASS00276, PASS00092, and PASS00742) obtained for master proteins encoded by chromosome 18. These proteins were measured in three types of biomaterial, including human plasma, liver samples, and HepG2 cells. The measurements were conducted according to Tier 3 (exploratory studies) guidelines using the double targeted strategy, which combines chromosome-centric approach with bottom-up SRM mass spectrometry.

A bell-shaped distribution histogram for master proteins encoded by chromosome 18 was observed revealing median of 10^8 copies per 1 μL of blood plasma and 10^5 copies per liver/HepG2 cell. The ascending portion of the curve reflects high- and medium-copied proteins, whereas the descending portion may be explained by either diminished proteome diversity in a biological sample or more probably the notion that the proteins cannot be detected due to low sensitivity of the analytical methods. Interestingly, after increasing the sensitivity of the analytical method from 10^{-14}M to 10^{-18} M by irreversible binding of analytes, 14 additional low-copied protein species (<105 copies per cell or per 1 μL of blood plasma) were gained and quantitatively measured, with at least two proteotypic peptides in each type of biomaterial. According to the results, there are much more high-abundant protein species in the plasma as compared with the liver or HepG2 cells. It is, therefore, likely that the difficulty in identifying ultralow-copied proteins in plasma is related to the high dynamic concentration range of plasma proteins.

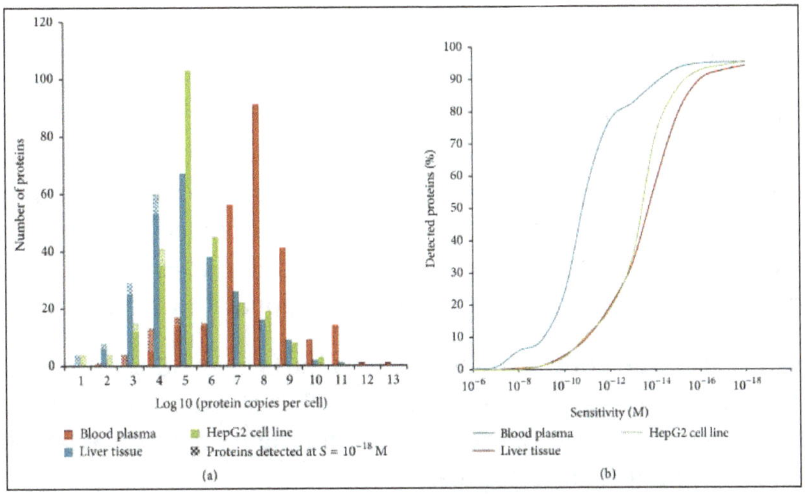

a) Distribution of the copy numbers of master proteins of chromosome 18 normalized per single HepG2/liver cell or 1 μL of plasma. (b) Share as a function of the detected proteins (in % to the total number of chromosome 18-coded proteins) and the analytical sensitivity.

To demonstrate the proteome depth, the number of copies of a master protein in a biosample was plotted depending on the sensitivity of proteomic technology. The proteome coverage was expressed as percent share of detected proteins to the total number of chromosome 18 genes, which was 276 according to neXtProt data. The distribution curve for the plasma proteins shifts left relative to the curves for the cells. The total number of detected protein species in liver and HepG2 cells increased relative to human blood plasma.

Future successes in human proteome exploration depend upon the ability to use bioinformatics methods to elucidate existing protein species and targeted MS analysis, high-throughput measurement, and high-performance algorithms for de novo assembly of protein sequences based on MS results. Furthermore, increasing the sensitivity of analytical technology will enable greater access to ultralow-copied proteins and expand opportunities for detection and analysis. In this context, theoretical prediction of the number of proteoforms (estimation of proteome width) and their distribution across the dynamic range (i.e., proteome depth) is ultimately required for planning the workload for the chromosome-centric Human Proteome Project.

Human Anatomy

Human anatomy is the scientific study of the body's structures. Some of these structures are very small and can only be observed and analyzed with the assistance of a microscope. Other larger structures can readily be seen, manipulated, measured, and weighed. The word "anatomy" comes from a Greek root that means "to cut apart." Human anatomy was first studied by observing the exterior of the body and observing the wounds of soldiers and other injuries. Later, physicians were allowed to dissect bodies of the dead to augment their knowledge. When a body is dissected, its structures are cut apart in order to observe their physical attributes and their relationships to one another. Human anatomy consists of two main divisions:

- Macroscopic or gross anatomy.
- Microscopic anatomy.

Let's begin by taking a look at macroscopic or gross anatomy. As the name suggests, this branch deals with large structures that are mostly seen with the naked eye. It describes where every human body structure is located (topography), similar to how a geographic map of an area shows all the landmarks in a particular perimeter. Not only that, but it also describes how the structures are connected to each other, their starting and end points, their layering, and so on. There are two fundamental approaches to studying gross anatomy: a regional and a systemic one.

Regional anatomy organizes the body into several body parts or regions: upper limbs, lower limbs, trunk (thorax, abdomen, pelvis, back), head, and neck. This approach divides teaching and learning into discrete regional didactic areas, each one containing its respective bones, joints, muscles, arteries, veins, nerves, lymphatics, and organs.

Upper Limb

The structures responsible for interacting with the environment, locomotion, weight bearing, and many more. The upper limb consists of four main parts: shoulder, arm, forearm, and hand. In turn, the mobility of the limb is provided by the shoulder, elbow, and wrist joints onto which various muscles act. However, their action depends on innervation, and their viability on proper nutrition and blood supply. Do you know the name of the vein punctured by the needle when you have your blood drawn? How about the name of the nerve that can lead to tingles if you lean on your elbows for a long time.

Lower Limb

The lower limb has four main parts, called the hip, thigh, leg, and foot. The flexibility is provided by the hip, knee, and ankle joints which allow you to kick, jump, squat, and shake it on the dance floor. The lower extremity contains some of the most powerful muscles in the human body, which are organized into various compartments. Significant vessels like the femoral artery and the longest nerve in the human body, the sciatic nerve, supply this limb.

Trunk and Back

The upper and lower limbs are attached to an anatomical structure called the trunk, commonly known as the torso. The trunk is composed of several regions called the thorax, abdomen, pelvis, and back. Running through the center of the back is the vertebral column which contains the spinal cord. Large back muscles such as the trapezius, latissimus dorsi, and rhomboids as well as deeper, smaller ones are attached to various points of the vertebral column. The musculature of the back helps you maintain your posture, bend your trunk, move your arms, shrug your shoulders, and much more. Large abdominal muscles, for example the rectus abdominis also contribute to the trunk. This is the famous 'six-pack' that many fitness enthusiasts strive for.

Thorax

The thorax is as complex on the inside as it is on the outside. Internally it consists of the thoracic cavity that, first and foremost, houses the lungs. These two vital organs are enveloped by membranes called the pleura and they are responsible for breathing. Altogether, the lungs occupy a surface area equivalent to the size of a tennis court. Sandwiched between the lungs is the mediastinum, a space that contains blood vessels, nerves, lymphatics, and most importantly, the heart. This vital organ is enclosed inside a sac called the pericardium and pumps 5 liters of blood every minute of your waking day through the entire body.

Abdomen and Pelvis

These two regions are often taught separately for didactic purposes, but their contents blend together into one large abdominopelvic cavity. Internally, it is lined by a membrane called the peritoneum which wraps around many structures, making them intraperitoneal. The ones located outside the membrane are named extraperitoneal. The largest organ system located here is the gastrointestinal tract. The intestines, which are mainly responsible for absorption, snail through these regions for a total of 7.5 meters, the equivalent of four human beings stacked upright on top of each other.

Four accessory organs that help the gastrointestinal tract to perform its functions are located inside the abdominopelvic cavity. These are the liver, gallbladder, pancreas, and spleen. They help especially with the digestion of proteins and fats, as well as metabolic processing.

It's easy to think that the abdomen and pelvis are overfilled with the gastrointestinal tract, but there's even more. Organs like the kidneys, ureters, urinary bladder, and female and male reproductive structures are also located here. They form entire systems that work in unison to ensure that you eliminate wastes, react to stressful or scary situations, and reproduce.

Your abdomen and pelvis are the home to some of the body's largest blood vessels. Since they supply major organs and even more distal body parts, they are large caliber structures transporting liters of blood. For example, if either the aorta or a renal artery ruptures during a traumatic event, the person would die in several minutes. Important nerves can also be found in these regions, controlling the activity of the abdominopelvic organs and allowing you to feel pain.

Head and Neck

In addition to the limbs, two more regions extend from the trunk that work in perfect harmony; a strong and mobile neck that supports a five-kilogram head, which also includes the brain. Vital nerves and blood vessels pass through the neck while traveling between the head and the rest of the body, so mastering these regions is important.

It consists of several bones joined together that form the bony skull, or cranium, parts of which enclose the brain and some that form the facial skeleton. The head has several associated structures, such as eyes, nose, ears, and mouth. They have a variety of functions, for instance, vision, smell, hearing, eating, and speaking, to name a few.

The neck serves as a passageway between the head and thorax. The nasal and oral cavities are continued by the pharynx, commonly called the throat. This muscular passage facilitates the movement of liquids, food, and air towards your windpipe (trachea) and food pipe (esophagus), respectively. In addition to the pharynx, the neck also houses many cartilages, muscles, organs, blood vessels, and nerves. Important structures include the larynx (voice box), thyroid gland, hyoid muscles, carotid arteries, jugular veins, and cervical plexus.

Neuroanatomy

Knowing the anatomy of each region of the human body is essential. However, how does the brain communicate with other regions, for instance the hand, to produce a movement or to sense objects? Through nerves, a concept explained by neuroanatomy. The nervous system controls every function of the human body. For example, it is involved in physiological processes like body temperature, voluntary movements, and higher-order thinking such as consciousness and emotional behaviour.

The nervous system has two structural divisions, central and peripheral. The central nervous system (CNS) consists of the brain and the spinal cord, which are protected by layers called meninges and bathed in cerebrospinal fluid. The brain is the master regulator of the body and has four main parts: cerebrum, subcortical structures, brainstem, and cerebellum. The cerebrum is divided into five lobes and forms the largest part of the human brain, being responsible for cognition.

However, every part of the brain is equally important. Do you know the main hero keeping a patient alive during a 'vegetative state' or a coma? It's solely the brainstem, as the cerebrum is dysfunctional.

The spinal cord is the continuation of the brainstem, travelling through the vertebral column. It consists of five regions called cervical, thoracic, lumbar, sacral, and coccyx. Spinal nerves emerge from the spinal cord through the vertebrae, carrying nerve impulses to and from the periphery. The brain and spinal cord communicate via neural pathways called tracts. Ascending tracts carry peripheral information up towards the brain, while descending tracts transport information back down from the brain.

The peripheral nervous system (PNS) refers to all neural tissue located outside the CNS. It consists of 12 pairs of cranial nerves, the 31 pairs of spinal nerves mentioned previously, and all their branches. The PNS reaches and innervates every single anatomical structure of the human body.

The regions of the human body are extremely complex. Starting from the top, the head allows you to gather information via sensorial structures, but not only, while the brain integrates and controls everything via the nervous system. The head rests on the neck, which provides a passageway for structures traveling to and from the thorax. Below the neck is the trunk, which is composed of the thoracic, abdominal, pelvic, and back regions. The trunk supports the body, facilitates movements, and protects various anatomical structures, such as internal organs, blood vessels, and nerves which are situated inside the respective cavities. Attached to it are two upper and lower limbs, which allow you to function as a human being, move, interact with your environment, and many more.

Systemic Anatomy

Systemic anatomy, the second branch of human anatomy, subdivides the body into discrete organ systems that work together towards a common goal or function. The ten systems are called integumentary, musculoskeletal (skeletal, muscular), nervous, endocrine, circulatory, respiratory, digestive, urinary, reproductive, and lymphatic.

In contrast to its previous counterpart, the systemic approach divides studying into areas dealing with specific functions rather than locations or proximity. This approach covers anatomy from a more physiological perspective, studying structures that accomplish one body function together. For example, the nervous system covers all nerves in the body, which span different regions from the brain to the lower limb.

Integumentary System

The integumentary system consists of the skin and its associated appendages, such as hair follicles, nails, sweat glands, and sebaceous glands. It is involved in a variety of functions, including protection, tactile and thermal sensation, as well as temperature regulation through sweating.

Musculoskeletal System

The 200 bones of the human body act as a scaffold, providing support, protection, facilitating locomotion, and even storing various cells and substances. The bones are the pulley systems onto which muscles act, the latter being capable of contracting and relaxing, ultimately producing movement.

Nervous System

Nerves are responsible for transporting electrical impulses, which permit communication between the brain, spinal cord, our senses, and every peripherical anatomical structure. This communication allows humans to interact with their environment, sense, feel emotions, think, and perform many other complex cognitive tasks.

Endocrine System

The endocrine system consists of glands that release substances called hormones into the bloodstream. These hormones allow chemical communication between anatomical structures, inducing various regulatory effects upon reaching their targets.

Circulatory System

The circulatory system is responsible for keeping us alive by providing oxygenated, arterial blood, to every part of the human body. The key player is the heart, an organ that pumps oxygenated blood into arteries, which then returns back to the heart via veins as deoxygenated blood.

Respiratory System

The main function of the respiratory system is to keep us alive via the inhalation of oxygen and elimination of carbon dioxide. The lungs and alveoli represent the site of gas exchange, which involve a series of convoluted air passages and membranes.

Digestive System

The digestive system is essentially a hollow system with two openings consisting of several organs. Food is ingested via the mouth, it is processed and absorbed inside the system, and the resulting solid wastes called feces are eliminated via the anus.

Urinary System

The urinary (excretory) system is the major filtering unit of the human body, being responsible for purifying the blood and eliminating wastes. The entire blood is continuously passed through the kidneys and the resulting unwanted or toxic substances are passed to the bladder, ultimately being eliminated via the urethra.

Reproductive System

The main responsibility of the reproductive system is to facilitate the generation of new offspring and to pass on our genes. The female system produces the eggs and nourishes the developing fetus until birth, while the male system synthesizes sperm and delivers it towards the egg to aid fertilization.

Lymphatic System

The lymphatic system is involved in the removal of interstitial fluid from tissues, the transport of absorbed fats after digestion, and in protection. Immune cells and substances travel via the lymphatic system, sampling lymph for any potential invaders and mounting an immune response if needed. Here are a lot of systems, so let's quickly summarize them. The integumentary system covers the entire body, protecting it from damage and regulating body temperature. The scaffold is provided by the skeletal system onto which the muscles act, ultimately facilitating movement. The nervous and endocrine systems are the main regulators, controlling the activity of almost everything via nerves and hormones, respectively. The respiratory and circulatory systems keep us alive by facilitating breathing and pumping blood around the body, while the digestive systems permits feeding. The resulting wastes are removed and excreted from the body by the urinary system. Last but not least, the reproductive system prevents humanity from going extinct, while the lymphatic system transports lymph and plays a role in the protection from microorganisms.

Microscopic Anatomy

Microscopic anatomy and histology are often used interchangeably, but they are quite distinct. Histology has a much broader scope, dealing with the structure and organization of tissues at all levels, from the intracellular components, to cells, and all the way to organs. In contrast, microscopic anatomy has a narrower scope, dealing only with 'microstructures' and the organization of tissues into organs. Let's use the example of the skeletal muscle motor unit to put it into context.

Microscopic anatomy describes skeletal muscle as consisting of fascicles and subsequent fibers, while neurons consisting of axons. However, histology explains much more, including the internal structure of the fibers, of each muscle cell, of the axons, as well as the appearance of the nuclei, and so on.

Human Physiology

Human Physiology studies the "nature" of the human body, nature in the sense of how structures at different levels work. Physiology focuses on function, or how structures at different levels work. The major systems covered in the study of human physiology are as follows:

- Circulatory system: Including the heart, the blood vessels, properties of the blood, and how circulation works in sickness and health.
- Digestive/excretory system: Charting the movement of solids from the mouth to the anus; this includes study of the spleen, liver, and pancreas, the conversion of food into fuel and its final exit from the body.
- Endocrine system: The study of endocrine hormones that carry signals throughout the organism, helping it to respond in concert. The principal endocrine glands – the pituitary, thyroid, adrenals, pancreas, parathyroids, and gonads – are a major focus, but nearly all organs release endocrine hormones.
- Immune system: The body's natural defense system is comprised of white blood cells, the thymus, and lymph systems. A complex array of receptors and molecules combine to protect the host from attacks by pathogens. Molecules such as antibodies and cytokines feature heavily.
- Integumentary system: The skin, hair, nails, sweat glands, and sebaceous glands (secreting an oily or waxy substance).
- Musculoskeletal system: The skeleton and muscles, tendons, ligaments, and cartilage. Bone marrow – where red blood cells are made – and how bones store calcium and phosphate are included.
- Nervous system: The central nervous system (brain and spinal cord) and the peripheral nervous system. Study of the nervous system includes research into the senses, memory, emotion, movement, and thought.
- Renal/urinary system: Including the kidneys, ureters, bladder, and urethra, this system removes water from the blood, produces urine, and carries away waste.
- Reproductive system: Consisting of the gonads and the sex organs. Study of this system also includes investigating the way a fetus is created and nurtured for 9 months.
- Respiratory system: Consisting of the nose, nasopharynx, trachea, and lungs. This system brings in oxygen and expels carbon dioxide and water.

Branches

Defense physiology investigates nature's natural defensive reactions.

There are a great number of disciplines that use the word physiology in their title. Below are some examples:

- Cell physiology: Studying the way cells work and interact; cell physiology mostly concentrates on membrane transport and neuron transmission.
- Systems physiology: This focuses on the computational and mathematical modeling of complex biological systems. It tries to describe the way individual cells or components of a system converge to respond as a whole. They often investigate metabolic networks and cell signaling.
- Evolutionary physiology: Studying the way systems, or parts of systems, have adapted and changed over multiple generations. Research topics cover a lot of ground including the role of behavior in evolution, sexual selection, and physiological changes in relation to geographic variation.
- Defense physiology: Changes that occur as a reaction to a potential threat, such as preparation for the fight-or-flight response.
- Exercise physiology: As the name suggests, this is the study of the physiology of physical exercise. This includes research into bioenergetics, biochemistry, cardiopulmonary function, biomechanics, hematology, skeletal muscle physiology, neuroendocrine function, and nervous system function.

Human Genetics

Human genetics is the study of the inheritance of characteristics by children from parents. Inheritance in humans does not differ in any fundamental way from that in other organisms.

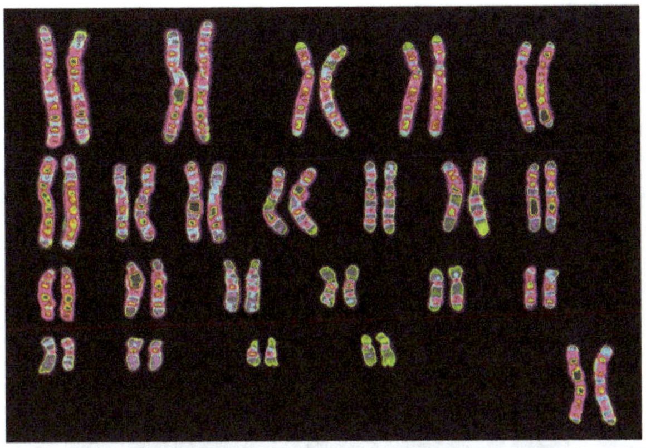

Hereditary information is contained in genes, which are carried on chromosomes.

The study of human heredity occupies a central position in genetics. Much of this interest stems from a basic desire to know who humans are and why they are as they are. At a more practical level, an understanding of human heredity is of critical importance in the prediction, diagnosis, and treatment of diseases that have a genetic component. The quest to determine the genetic basis of human health has given rise to the field of medical genetics. In general, medicine has given focus and purpose to human genetics, so the terms medical genetics and human genetics are often considered synonymous.

The Human Chromosomes

A new era in cytogenetics, the field of investigation concerned with studies of the chromosomes, began in 1956 with the discovery by Jo Hin Tjio and Albert Levan that human somatic cells contain 23 pairs of chromosomes. Since that time the field has advanced with amazing rapidity and has demonstrated that human chromosome aberrations rank as major causes of fetal death and of tragic human diseases, many of which are accompanied by intellectual disability. Since the chromosomes can be delineated only during mitosis, it is necessary to examine material in which there are many dividing cells. This can usually be accomplished by culturing cells from the blood or skin, since only the bone marrow cells (not readily sampled except during serious bone marrow disease such as leukemia) have sufficient mitoses in the absence of artificial culture. After growth, the cells are fixed on slides and then stained with a variety of DNA-specific stains that permit the delineation and identification of the chromosomes. The Denver system of chromosome classification, established in 1959, identified the chromosomes by their length and the position of the centromeres. Since then the method has been improved by the use of special staining techniques that impart unique light and dark bands to each chromosome. These bands permit the identification of chromosomal regions that are duplicated, missing, or transposed to other chromosomes.

Micrographs showing the karyotypes (i.e., the physical appearance of the chromosome) of a male and a female have been produced. In a typical micrograph the 46 human chromosomes (the diploid number) are arranged in homologous pairs, each consisting of one maternally derived and one paternally derived member. The chromosomes are all numbered except for the X and the Y chromosomes, which are the sex chromosomes. In humans, as in all mammals, the normal female has two X chromosomes and the normal male has one X chromosome and one Y chromosome. The female is thus the homogametic sex, as all her gametes normally have one X chromosome. The male is heterogametic, as he produces two types of gametes—one type containing an X chromosome and the other containing a Y chromosome. There is good evidence that the Y chromosome in humans, unlike that in Drosophila, is necessary (but not sufficient) for maleness.

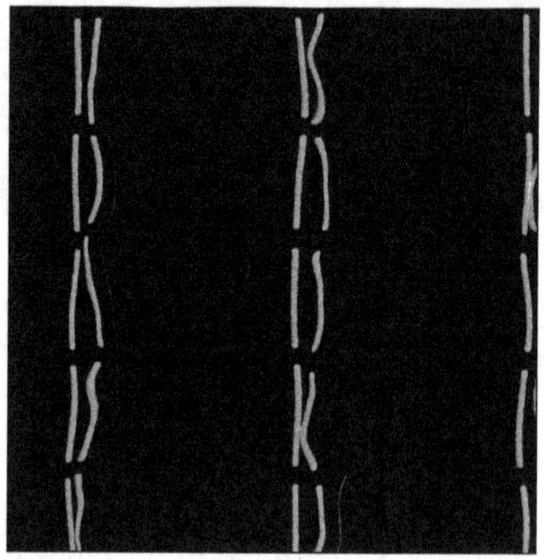

Chromosome: Strands of human chromosomes.

Fertilization, Sex Determination and Differentiation

A human individual arises through the union of two cells, an egg from the mother and a sperm from the father. Human egg cells are barely visible to the naked eye. They are shed, usually one at a time, from the ovary into the oviducts (fallopian tubes), through which they pass into the uterus. Fertilization, the penetration of an egg by a sperm, occurs in the oviducts. This is the main event of sexual reproduction and determines the genetic constitution of the new individual.

Human sex determination is a genetic process that depends basically on the presence of the Y chromosome in the fertilized egg. This chromosome stimulates a change in the undifferentiated gonad into that of the male (a testicle). The gonadal action of the Y chromosome is mediated by a gene located near the centromere; this gene codes for the production of a cell surface molecule called the H-Y antigen. Further development of the anatomic structures, both internal and external, that are associated with maleness is controlled by hormones produced by the testicle. The sex of an individual can be thought of in three different contexts: chromosomal sex, gonadal sex, and anatomic sex. Discrepancies between these, especially the latter two, result in the development of individuals with ambiguous sex, often called hermaphrodites. Homosexuality is unrelated to the above sex-determining factors. It is of interest that in the absence of a male gonad (testicle) the internal and external sex anatomy is always female, even in the absence of a female ovary. A female without ovaries will, of course, be infertile and will not experience any of the female developmental changes normally associated with puberty. Such a female will often have Turner syndrome.

If X-containing and Y-containing sperm are produced in equal numbers, then according to simple chance one would expect the sex ratio at conception (fertilization) to be half boys and half girls, or 1 : 1. Direct observation of sex ratios among newly fertilized human eggs is not yet feasible, and sex-ratio data are usually collected at the time of birth. In almost all human populations of newborns, there is a slight excess of males; about 106 boys are born for every 100 girls. Throughout life, however, there is a slightly greater mortality of males; this slowly alters the sex ratio until, beyond the age of about 50 years, there is an excess of females. Studies indicate that male embryos suffer a relatively greater degree of prenatal mortality, so the sex ratio at conception might be expected to favour males even more than the 106 : 100 ratio observed at birth would suggest. Firm explanations for the apparent excess of male conceptions have not been established; it is possible that Y-containing sperm survive better within the female reproductive tract, or they may be a little more successful in reaching the egg in order to fertilize it. In any case, the sex differences are small, the statistical expectation for a boy (or girl) at any single birth still being close to one out of two.

During gestation—the period of nine months between fertilization and the birth of the infant—a remarkable series of developmental changes occur. Through the process of mitosis, the total number of cells changes from 1 (the fertilized egg) to about 2 × 1011. In addition, these cells differentiate into hundreds of different types with specific functions (liver cells, nerve cells, muscle cells, etc.). A multitude of regulatory processes, both genetically and environmentally controlled, accomplish this differentiation. Elucidation of the exquisite timing of these processes remains one of the great challenges of human biology.

Immunogenetics

Immunity is the ability of an individual to recognize the "self" molecules that make up one's own body and to distinguish them from such "oneself" molecules as those found in infectious microorganisms and toxins. This process has a prominent genetic component. Knowledge of the genetic and molecular basis of the mammalian immune system has increased in parallel with the explosive advances made in somatic cell and molecular genetics.

There are two major components of the immune system, both originating from the same precursor "stem" cells. The bursa component provides B lymphocytes, a class of white blood cells that, when appropriately stimulated, differentiate into plasma cells. These latter cells produce circulating soluble proteins called antibodies or immunoglobulins. Antibodies are produced in response to substances called antigens, most of which are foreign proteins or polysaccharides. An antibody molecule can recognize a specific antigen, combine with it, and initiate its destruction. This so-called humoral immunity is accomplished through a complicated series of interactions with other molecules and cells; some of these interactions are mediated by another group of lymphocytes, the T lymphocytes, which are derived from the thymus gland. Once a B lymphocyte has been exposed to a specific antigen, it "remembers" the contact so that future exposure will cause an accelerated and magnified immune reaction. This is a manifestation of what has been called immunological memory.

The thymus component of the immune system centres on the thymus-derived T lymphocytes. In addition to regulating the B cells in producing humoral immunity, the T cells also directly attack cells that display foreign antigens. This process, called cellular immunity, is of great importance in protecting the body against a variety of viruses as well as cancer cells. Cellular immunity is also the chief cause of the rejection of organ transplants. The T lymphocytes provide a complex network consisting of a series of helper cells (which are antigen-specific), amplifier cells, suppressor cells, and cytotoxic (killer) cells, all of which are important in immune regulation.

The Genetics of Antibody Formation

One of the central problems in understanding the genetics of the immune system has been in explaining the genetic regulation of antibody production. Immunobiologists have demonstrated that the system can produce well over one million specific antibodies, each corresponding to a particular antigen. It would be difficult to envisage that each antibody is encoded by a separate gene; such an arrangement would require a disproportionate share of the entire human genome. Recombinant DNA analysis has illuminated the mechanisms by which a limited number of immunoglobulin genes can encode this vast number of antibodies.

Each antibody molecule consists of several different polypeptide chains—the light chains (L) and the longer heavy chains (H). The latter determine to which of five different classes (IgM, IgG, IgA, IgD, or IgE) an immunoglobulin belongs. Both the L and H chains are unique among proteins in that they contain constant and variable parts. The constant parts have relatively identical amino acid sequences in any given antibody. The variable parts, on the other hand, have different amino acid sequences in each antibody molecule. It is the variable parts, then, that determine the specificity of the antibody.

Recombinant DNA studies of immunoglobulin genes in mice have revealed that the light-chain genes are encoded in four separate parts in germ-line DNA: a leader segment (L), a variable segment (V), a joining segment (J), and a constant segment (C). These segments are widely separated in the DNA of an embryonic cell, but in a mature B lymphocyte they are found in relative proximity (albeit separated by introns). The mouse has more than 200 light-chain variable region genes, only one of which will be incorporated into the proximal sequence that codes for the antibody production in a given B lymphocyte. Antibody diversity is greatly enhanced by this system, as the V and J segments rearrange and assort randomly in each B-lymphocyte precursor cell. The mechanisms by which this DNA rearrangement takes place are not clear, but transposons are undoubtedly involved. Similar combinatorial processes take place in the genes that code for the heavy chains; furthermore, both the light-chain and heavy-chain genes can undergo somatic mutations to create new antibody-coding sequences. The net effect of these combinatorial and mutational processes enables the coding of millions of specific antibody molecules from a limited number of genes. It should be stressed, however, that each B lymphocyte can produce only one antibody. It is the B lymphocyte population as a whole that produces the tremendous variety of antibodies in humans and other mammals.

Plasma cell tumours (myelomas) have made it possible to study individual antibodies, since these tumours, which are descendants of a single plasma cell, produce one antibody in abundance. Another method of obtaining large amounts of a specific antibody is by fusing a B lymphocyte with a rapidly growing cancer cell. The resultant hybrid cell, known as a hybridoma, multiplies rapidly in culture. Since the antibodies obtained from hybridomas are produced by clones derived from a single lymphocyte, they are called monoclonal antibodies.

The Genetics of Cellular Immunity

Cellular immunity is mediated by T lymphocytes that can recognize infected body cells, cancer cells, and the cells of a foreign transplant. The control of cellular immune reactions is provided by a linked group of genes, known as the major histocompatibility complex (MHC). These genes code for the major histocompatibility antigens, which are found on the surface of almost all nucleated somatic cells. The major histocompatibility antigens were first discovered on the leukocytes (white blood cells) and are therefore usually referred to as the HLA (human leukocyte group A) antigens.

The advent of the transplantation of human organs in the 1950s made the question of tissue compatibility between donor and recipient of vital importance, and it was in this context that the HLA antigens and the MHC were elucidated. Investigators found that the MHC resides on the short arm of chromosome 6, on four closely associated sites designated HLA-A, HLA-B, HLA-C, and HLA-D. Each locus is highly polymorphic; i.e., each is represented by a great many alleles within the human gene pool. These alleles, like those of the ABO blood group system, are expressed in co-dominant fashion. Because of the large number of alleles at each HLA locus, there is an extremely low probability of any two individuals (other than siblings) having identical HLA genotypes. (Since a person inherits one chromosome 6 from each parent, siblings have a 25 percent probability of having received the same paternal and maternal chromosomes 6 and thus of being HLA matched).

Although HLA antigens are largely responsible for the rejection of organ transplants, it is obvious that the MHC did not evolve to prevent the transfer of organs from one person to another. Indeed, information obtained from the histocompatibility complex in the mouse (which is very similar in its genetic organization to that of the human) suggests that a primary function of the HLA antigens is to regulate the number of specific cytotoxic T killer cells, which have the ability to destroy virus-infected cells and cancer cells.

The Genetics of Human Blood

More is known about the genetics of the blood than about any other human tissue. One reason for this is that blood samples can be easily secured and subjected to biochemical analysis without harm or major discomfort to the person being tested. Perhaps a more cogent reason is that many chemical properties of human blood display relatively simple patterns of inheritance.

Blood Types

Certain chemical substances within the red blood cells (such as the ABO and MN substances may serve as antigens. When cells that contain specific antigens are introduced into the body of an experimental animal such as a rabbit, the animal responds by producing antibodies in its own blood.

In addition to the ABO and MN systems, geneticists have identified about 14 blood-type gene systems associated with other chromosomal locations. The best known of these is the Rh system. The Rh antigens are of particular importance in human medicine. Curiously, however, their existence was discovered in monkeys. When blood from the rhesus monkey (hence the designation Rh) is injected into rabbits, the rabbits produce so-called Rh antibodies that will agglutinate not only the red blood cells of the monkey but the cells of a large proportion of human beings as well. Some people (Rh-negative individuals), however, lack the Rh antigen; the proportion of such persons varies from one human population to another. Akin to data concerning the ABO system, the evidence for Rh genes indicates that only a single chromosome locus (called r) is involved and is located on chromosome 1. At least 35 Rh alleles are known for the r location; basically the Rh-negative condition is recessive.

A medical problem may arise when a woman who is Rh-negative carries a fetus that is Rh-positive. The first such child may have no difficulty, but later similar pregnancies may produce severely anemic newborn infants. Exposure to the red blood cells of the first Rh-positive fetus appears to immunize the Rh-negative mother, that is, she develops antibodies that may produce permanent (sometimes fatal) brain damage in any subsequent Rh-positive fetus. Damage arises from the scarcity of oxygen reaching the fetal brain because of the severe destruction of red blood cells. Measures are available for avoiding the severe effects of Rh incompatibility by transfusions to the fetus within the uterus; however, genetic counselling before conception is helpful so that the mother can receive Rh immunoglobulin immediately after her first and any subsequent pregnancies involving an Rh-positive fetus. This immunoglobulin effectively destroys the fetal red blood cells before the mother's immune system is stimulated. The mother thus avoids becoming actively immunized against the Rh antigen and will not produce antibodies that could attack the red blood cells of a future Rh-positive fetus.

Serum Proteins

Human serum, the fluid portion of the blood that remains after clotting, contains various proteins that have been shown to be under genetic control. Study of genetic influences has flourished since the development of precise methods for separating and identifying serum proteins. These move at different rates under the impetus of an electrical field (electrophoresis), as do proteins from many other sources (e.g., muscle or nerve). Since the composition of a protein is specified by the structure of its corresponding gene, biochemical studies based on electrophoresis permit direct study of tissue substances that are only a metabolic step or two away from the genes themselves.

Electrophoretic studies have revealed that at least one-third of the human serum proteins occur in variant forms. Many of the serum proteins are polymorphic, occurring as two or more variants with a frequency of not less than 1 percent each in a population. Patterns of polymorphic serum protein variants have been used to determine whether twins are identical (as in assessing compatibility for organ transplants) or whether two individuals are related (as in resolving paternity suits). Whether the different forms have a selective advantage is not generally known.

Much attention in the genetics of substances in the blood has been centred on serum proteins called haptoglobins, transferrins (which transport iron), and gamma globulins (a number of which are known to immunize against infectious diseases). Haptoglobins appear to relate to two common alleles at a single chromosome locus; the mode of inheritance of the other two seems more complicated, about 18 kinds of transferrins having been described. Like blood-cell antigen genes, serum-protein genes are distributed worldwide in the human population in a way that permits their use in tracing the origin and migration of different groups of people.

Hemoglobin

Hundreds of variants of hemoglobin have been identified by electrophoresis, but relatively few are frequent enough to be called polymorphisms. Of the polymorphisms, the alleles for sickle-cell and thalassemia hemoglobins produce serious disease in homozygotes, whereas others (hemoglobins C, D, and E) do not. The sickle-cell polymorphism confers a selective advantage on the heterozygote living in a malarial environment; the thalassemia polymorphism provides a similar advantage.

Influence of the Environment

Gene expression occurs only after modification by the environment. A good example is the recessively inherited disease called galactosemia, in which the enzyme necessary for the metabolism of galactose—a component of milk sugar—is defective. The sole source of galactose in the infant's diet is milk, which in this instance is toxic. The treatment of this most serious disease in the neonate is to remove all natural forms of milk from the diet (environmental manipulation) and to substitute a synthetic milk lacking galactose. The infant will then develop normally but will never be able to tolerate foods containing lactose. If milk was not a major part of the infant's diet, however, the mutant gene would never be able to express itself, and galactosemia would be unknown.

Another way of saying this is that no trait can exist or become actual without an environmental contribution. Thus, the old question of which is more important, heredity or environment, is without meaning. Both nature (heredity) and nurture (environment) are always important for every human attribute.

But this is not to say that the separate contributions of heredity and environment are equivalent for each characteristic. Dark pigmentation of the iris of the eye, for example, is under hereditary control in that one or more genes specify the synthesis and deposition in the iris of the pigment (melanin). This is one characteristic that is relatively independent of such environmental factors as diet or climate; thus, individual differences in eye colour tend to be largely attributable to hereditary factors rather than to ordinary environmental change.

On the other hand, it is unwarranted to assume that other traits (such as height, weight, or intelligence) are as little affected by environment as is eye colour. It is very easy to gather information that tall parents tend, on the average, to have tall children (and that short parents tend to produce short children), properly indicating a hereditary contribution to height. Nevertheless, it is equally manifest that growth can be stunted in the environmental absence of adequate nutrition. The dilemma arises that only the combined, final result of this nature-nurture interaction can be directly observed. There is no accurate way (in the case of a single individual) to gauge the separate contributions of heredity and environment to such a characteristic as height. An inferential way out of this dilemma is provided by studies of twins.

Fraternal Twins

Usually a fertile human female produces a single egg about once a month. Should fertilization occur (a zygote is formed), growth of the individual child normally proceeds after the fertilized egg has become implanted in the wall of the uterus (womb). In the unusual circumstance that two unfertilized eggs are simultaneously released by the ovaries, each egg may be fertilized by a different sperm cell at about the same time, become implanted, and grow, to result in the birth of twins.

Twins formed from separate eggs and different sperm cells can be of the same or of either sex. No matter what their sex, they are designated as fraternal twins. This terminology is used to emphasize that fraternal twins are genetically no more alike than are siblings (brothers or sisters) born years apart. Basically they differ from ordinary siblings only in having grown side by side in the womb and in having been born at approximately the same time.

Identical Twins

In a major nonfraternal type of twinning, only one egg is fertilized, but during the cleavage of this single zygote into two cells, the resulting pair somehow becomes separated. Each of the two cells may implant in the uterus separately and grow into a complete, whole individual. In laboratory studies with the zygotes of many animal species, it has been found that in the two-cell stage (and later) a portion of the embryo, if separated under the microscope by the experimenter, may develop into a perfect, whole individual. Such splitting occurs spontaneously at the four-cell stage in some organisms (e.g., the armadillo) and has been accomplished experimentally with the embryos of salamanders, among others.

The net result of splitting at an early embryonic stage may be to produce so-called identical twins. Since such twins derive from the same fertilized egg, the hereditary material from which they originate is absolutely identical in every way, down to the last gene locus. While developmental and genetic differences between one "identical" twin and another still may arise through a number of processes (e.g., mutation), these twins are always found to be of the same sex. They are often breathtakingly similar in appearance, frequently down to very fine anatomic and biochemical details (although their fingerprints are differentiable).

Diagnosis of Twin Types

Since the initial event in the mother's body (either splitting of a single egg or two separate fertilizations) is not observed directly, inferential means are employed for diagnosing a set of twins as fraternal or identical. The birth of fraternal twins is frequently characterized by the passage of two separate afterbirths. In many instances, identical twins are followed by only a single afterbirth, but exceptions to this phenomenon are so common that this is not a reliable method of diagnosis.

The most trustworthy method for inferring twin type is based on the determination of genetic similarity. By selecting those traits that display the least variation attributable to environmental influences (such as eye colour and blood types), it is feasible, if enough separate chromosome loci are considered, to make the diagnosis of twin type with high confidence. HLA antigens, which , are very polymorphic, have become most useful in this regard.

Inferences from Twin Studies

Metric (Quantitative) Traits

By measuring the heights of a large number of ordinary siblings (brothers and sisters) and of twin pairs, it may be shown that the average difference between identical twins is less than half the difference for all other siblings. Any average differences between groups of identical twins are attributable with considerable confidence to the environment. Thus, since the sample of identical twins who were reared apart (in different homes) differed little in height from identicals who were raised together, it appears that environmental-genetic influences on that trait tended to be similar for both groups.

Yet, the data for like-sexed fraternal twins reveal a much greater average difference in height (about the same as that found between ordinary siblings reared in the same home at different ages). Apparently the fraternal twins were more dissimilar than identicals (even though reared together) because the fraternals differed more from each other in genotype. This emphasizes the great genetic similarity between identicals. Such studies can be particularly enlightening when the effects of individual genes are obscured or distorted by the influence of environmental factors on quantitative (measurable) traits (e.g., height, weight, and intelligence).

Any trait that can be objectively measured in identical and fraternal twins can be scrutinized for the particular combination of hereditary and environmental influences that impinge upon it. The effect of environment on identical twins reared apart is suggested by their relatively great average difference in body weight as compared with identical twins reared together. Weight appears to be more strongly modified by environmental variables than is height.

Study of comparable characteristics among farm animals and plants suggests that such quantitative human traits as height and weight are affected by allelic differences at a number of chromosome locations—that they are not simply affected by genes at a single locus. Investigation of these gene systems with multiple locations (polygenic systems) is carried out largely through selective-breeding experiments among large groups of plants and lower animals. Human beings select their mates in a much freer fashion, of course, and polygenic studies among people are thus severely limited.

Intelligence is a very complex human trait, the genetics of which has been a subject of controversy for some time. Much of the controversy arises from the fact that intelligence is so difficult to define. Information has been based almost entirely on scores on standardized IQ tests constructed by psychologists; in general, such tests do not take into account cultural, environmental, and educational differences. As a result, the working definition of intelligence has been "the general factor common to a large number of diverse cognitive (IQ) tests." Even roughly measured as IQ, intelligence shows a strong contribution from the environment. Fraternal twins, however, show relatively great dissimilarity in IQ, suggesting an important contribution from heredity as well. In fact, it has been estimated that, on the average, between 60 and 80 percent of the variance in IQ test scores could be genetic. It is important to note that intelligence is polygenically inherited and that it has the highest degree of assortative mating of any trait; in other words, people tend to mate with people having similar IQs. Moreover, twin studies involving psychological traits should be viewed with caution; for example, since identical twins tend to be singled out for special attention, their environment should not be considered equivalent even to that of other children raised in their own family.

Since the time of Galton, generalizations have been repeatedly made about racial differences in intelligence, with claims of genetic superiority of some races over others. These generalizations fail to recognize that races are composed of individuals, each of whom has a unique genotype made up by genes shared with other humans, and that the sources of intraracial variation are more numerous than those producing interracial differences.

Other Traits

For traits of a more qualitative (all-or-none) nature, the twin method can also be used in efforts to assess the degree of hereditary contribution. Such investigations are based on an examination of cases in which at least one member of the twin pair shows the trait. It was found in one study, for example, that in about 80 percent of all identical twin pairs in which one twin shows symptoms of the psychiatric disorder called schizophrenia, the other member of the pair also shows the symptoms; that is, the two are concordant for the schizophrenic trait. In the remaining 20 percent, the twins are discordant; that is, one lacks the trait. Since identical twins often have similar environments, this information by itself does not distinguish between the effects of heredity and environment. When pairs of like-sexed fraternal twins reared together are studied, however, the degree of concordance for schizophrenia is very much lower—only about 15 percent.

Schizophrenia thus clearly develops much more easily in some genotypes than in others; this indicates a strong hereditary predisposition to the development of the trait. Schizophrenia also serves as a good example of the influence of environmental factors, since concordance for the condition does not appear in 100 percent of identical twins.

Studies of concordance and discordance between identical and fraternal twins have been carried out for many other human characteristics. It has, for example, been known for many years that tuberculosis is a bacterial infection of environmental origin. Yet identical twins raised in the same home show concordance for the disease far more often than do fraternal twins. This finding seems to be explained by the high degree of genetic similarity between the identical twins. While the tuberculosis germ is not inherited, heredity does seem to make one more (or less) susceptible to this particular infection. Thus, the genes of one individual may provide the chemical basis for susceptibility to a disease, while the genes of another may fail to do so.

Indeed, there seem to be genetic differences between disease germs themselves that result in differences in their virulence. Thus, whether a genetically susceptible person actually develops a disease also depends in part on the heredity of the particular strain of bacteria or virus with which he or she must cope. Consequently, unless environmental factors such as these are adequately evaluated, the conclusions drawn from susceptibility studies can be unfortunately misleading.

References

- Human-body, science: britannica.com, Retrieved 05, Jan 2020
- How-genes-work, protein, genetics: medlineplus.gov, Retrieved 20, Jan 2020
- The-human-anatomy: kenhub.com, Retrieved 25, Feb 2020
- What-is-physiology: medicalnewstoday, Retrieved 10, Feb 2020
- Human-genetics, science: britannica.com, Retrieved 03, March 2020

Chapter 4

Biology of Plants

Botany is the scientific study of plants, their structure, growth, development, classification, diseases and environmental impacts. Considering the fundamental role of plants in human survival, the importance of botany has increased significantly. Plant anatomy, breeding, propagation, morphology, ecology, physiology and pathology are the important topics covered in this chapter.

Botany

Botany, also called plant science(s), plant biology or phytology, is the science of plant life and a branch of biology. A botanist or plant scientist is a scientist who specialises in this field. Traditionally, botany has also included the study of fungi and algae by mycologists and phycologists respectively, with the study of these three groups of organisms remaining within the sphere of interest of the International Botanical Congress. Nowadays, botanists study approximately 400,000 species of living organisms of which some 260,000 species are vascular plants and about 248,000 are flowering plants.

The fruit of *Myristica fragrans*, a species native to Indonesia, is the source of two valuable spices, the red aril (mace) enclosing the dark brown nutmeg.

Botany originated in prehistory as herbalism with the efforts of early humans to identify – and later cultivate – edible, medicinal and poisonous plants, making it one of the oldest branches of science. Medieval physic gardens, often attached to monasteries, contained plants of medical importance. They were forerunners of the first botanical gardens attached to universities, founded from the 1540s onwards. One of the earliest was the Padua botanical garden. These gardens facilitated the academic study of plants. Efforts to catalogue and describe their collections were the beginnings of plant taxonomy, and led in 1753 to the binomial system of Carl Linnaeus that remains in use to this day.

In the 19th and 20th centuries, new techniques were developed for the study of plants, including methods of optical microscopy and live cell imaging, electron microscopy, analysis of chromosome number, plant chemistry and the structure and function of enzymes and other proteins. In the last two decades of the 20th century, botanists exploited the techniques of molecular genetic analysis, including genomics and proteomics and DNA sequences to classify plants more accurately.

Modern botany is a broad, multidisciplinary subject with inputs from most other areas of science and technology. Research topics include the study of plant structure, growth and differentiation, reproduction, biochemistry and primary metabolism, chemical products, development, diseases, evolutionary relationships, systematics, and plant taxonomy. Dominant themes in 21st century plant science are molecular genetics and epigenetics, which are the mechanisms and control of gene expression during differentiation of plant cells and tissues. Botanical research has diverse applications in providing staple foods, materials such as timber, oil, rubber, fibre and drugs, in modern horticulture, agriculture and forestry, plant propagation, breeding and genetic modification, in the synthesis of chemicals and raw materials for construction and energy production, in environmental management, and the maintenance of biodiversity.

History

Early Botany

Botany originated as herbalism, the study and use of plants for their medicinal properties. Many records of the Holocene period date early botanical knowledge as far back as 10,000 years ago. This early unrecorded knowledge of plants was discovered in ancient sites of human occupation within Tennessee, which make up much of the Cherokee land today. The early recorded history of botany includes many ancient writings and plant classifications. Examples of early botanical works have been found in ancient texts from India dating back to before 1100 BC, in archaic Avestan writings, and in works from China before it was unified in 221 BC.

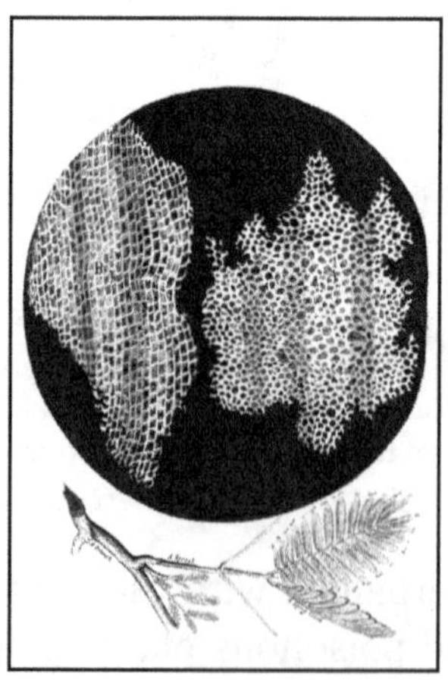

An engraving of the cells of cork, from Robert Hooke's *Micrographia,* 1665

Modern botany traces its roots back to Ancient Greece specifically to Theophrastus (c. 371–287 BC), a student of Aristotle who invented and described many of its principles and is widely regarded in the scientific community as the "Father of Botany". His major works, *Enquiry into Plants* and *On the Causes of Plants*, constitute the most important contributions to botanical science until the Middle Ages, almost seventeen centuries later.

Another work from Ancient Greece that made an early impact on botany is *De Materia Medica*, a five-volume encyclopedia about herbal medicine written in the middle of the first century by Greek physician and pharmacologist Pedanius Dioscorides. *De Materia Medica* was widely read for more than 1,500 years. Important contributions from the medieval Muslim world include Ibn Wahshiyya's *Nabatean Agriculture*, Abū Ḥanīfa Dīnawarī's (828–896) the *Book of Plants*, and Ibn Bassal's *The Classification of Soils*. In the early 13th century, Abu al-Abbas al-Nabati, and Ibn al-Baitar (d. 1248) wrote on botany in a systematic and scientific manner.

In the mid-16th century, "botanical gardens" were founded in a number of Italian universities – the Padua botanical garden in 1545 is usually considered to be the first which is still in its original location. These gardens continued the practical value of earlier "physic gardens", often associated with monasteries, in which plants were cultivated for medical use. They supported the growth of botany as an academic subject. Lectures were given about the plants grown in the gardens and their medical uses demonstrated. Botanical gardens came much later to northern Europe; the first in England was the University of Oxford Botanic Garden in 1621. Throughout this period, botany remained firmly subordinate to medicine.

German physician Leonhart Fuchs (1501–1566) was one of "the three German fathers of botany", along with theologian Otto Brunfels (1489–1534) and physician Hieronymus Bock (1498–1554) (also called Hieronymus Tragus). Fuchs and Brunfels broke away from the tradition of copying earlier works to make original observations of their own. Bock created his own system of plant classification.

Physician Valerius Cordus (1515–1544) authored a botanically and pharmacologically important herbal *Historia Plantarum* in 1544 and a pharmacopoeia of lasting importance, the *Dispensatorium* in 1546. Naturalist Conrad von Gesner (1516–1565) and herbalist John Gerard (1545–c. 1611) published herbals covering the medicinal uses of plants. Naturalist Ulisse Aldrovandi (1522–1605) was considered the *father of natural history*, which included the study of plants. In 1665, using an early microscope, Polymath Robert Hooke discovered cells, a term he coined, in cork, and a short time later in living plant tissue.

Early Modern Botany

During the 18th century, systems of plant identification were developed comparable to dichotomous keys, where unidentified plants are placed into taxonomic groups (e.g. family, genus and species) by making a series of choices between pairs of characters. The choice and sequence of the characters may be artificial in keys designed purely for identification (diagnostic keys) or more closely related to the natural or phyletic order of the taxa in synoptic keys. By the 18th century, new plants for study were arriving in Europe in increasing numbers from newly discovered countries and the European colonies worldwide. In 1753 Carl von Linné (Carl Linnaeus) published his Species Plantarum, a hierarchical classification of plant species that remains the reference point for modern botanical nomenclature. This established a standardised binomial or two-part naming scheme where the first name represented the genus and the second identified the species within the genus. For the purposes of identification, Linnaeus's *Systema Sexuale* classified plants into 24 groups according to the number of their male sexual organs. The 24th group, *Cryptogamia*, included all plants with concealed reproductive parts, mosses, liverworts, ferns, algae and fungi.

The Linnaean Garden of Linnaeus' residence in Uppsala, Sweden, was planted according to his *Systema sexuale*.

Increasing knowledge of plant anatomy, morphology and life cycles led to the realisation that there were more natural affinities between plants than the artificial sexual system of Linnaeus. Adanson (1763), de Jussieu (1789), and Candolle (1819) all proposed various alternative natural systems of classification that grouped plants using a wider range of shared characters and were widely followed. The Candollean system reflected his ideas of the progression of morphological complexity and the later classification by Bentham and Hooker, which was influential until the mid-19th century, was influenced by Candolle's approach. Darwin's publication of the *Origin of Species* in 1859 and his concept of common descent required modifications to the Candollean system to reflect evolutionary relationships as distinct from mere morphological similarity.

Botany was greatly stimulated by the appearance of the first "modern" textbook, Matthias Schleiden's *Grundzüge der Wissenschaftlichen Botanik*, published in English in 1849 as *Principles of Scientific Botany*. Schleiden was a microscopist and an early plant anatomist who co-founded the cell theory with Theodor Schwann and Rudolf Virchow and was among the first to grasp the significance of the cell nucleus that had been described by Robert Brown in 1831. In 1855, Adolf Fick formulated Fick's laws that enabled the calculation of the rates of molecular diffusion in biological systems.

Echeveria glauca in a Connecticut greenhouse. Botany uses Latin names for identification, here, the specific name *glauca* means blue.

Modern Botany

Building upon the gene-chromosome theory of heredity that originated with Gregor Mendel (1822–1884), August Weismann (1834–1914) proved that inheritance only takes place through gametes. No other cells can pass on inherited characters. The work of Katherine Esau (1898–1997) on plant anatomy is still a major foundation of modern botany. Her books *Plant Anatomy* and *Anatomy of Seed Plants* have been key plant structural biology texts for more than half a century.

Biologist and statistician Ronald Fisher.

Micropropagation of transgenic plants.

The discipline of plant ecology was pioneered in the late 19th century by botanists such as Eugenius Warming, who produced the hypothesis that plants form communities, and his mentor and successor Christen C. Raunkiær whose system for describing plant life forms is still in use today. The concept that the composition of plant communities such as temperate broadleaf forest changes by a process of ecological succession was developed by Henry Chandler Cowles, Arthur Tansley and Frederic Clements. Clements is credited with the idea of climax vegetation as the most complex vegetation that an environment can support and Tansley introduced the concept of ecosystems to biology. Building on the extensive earlier work of Alphonse de Candolle, Nikolai Vavilov (1887–1943) produced accounts of the biogeography, centres of origin, and evolutionary history of economic plants.

Particularly since the mid-1960s there have been advances in understanding of the physics of plant physiological processes such as transpiration (the transport of water within plant tissues), the temperature dependence of rates of water evaporation from the leaf surface and the molecular diffusion of water vapour and carbon dioxide through stomatal apertures. These developments, coupled with new methods for measuring the size of stomatal apertures, and the rate of photosynthesis have enabled precise description of the rates of gas exchange between plants and the atmosphere. Innovations in statistical analysis by Ronald Fisher, Frank Yates and others at Rothamsted Experimental Station facilitated rational experimental design and data analysis in botanical research. The discovery and identification of the auxin plant hormones by Kenneth V. Thimann in 1948 enabled regulation of plant growth by externally applied chemicals. Frederick Campion Steward pioneered techniques of micropropagation and plant tissue culture controlled by plant hormones. The synthetic auxin 2,4-Dichlorophenoxyacetic acid or 2,4-D was one of the first commercial synthetic herbicides.

20th century developments in plant biochemistry have been driven by modern techniques of organic chemical analysis, such as spectroscopy, chromatography and electrophoresis. With the rise of the related molecular-scale biological approaches of molecular biology, genomics, proteomics and metabolomics, the relationship between the plant genome and most aspects of the biochemistry, physiology, morphology and behaviour of plants can be subjected to detailed experimental analysis. The concept originally stated by Gottlieb Haberlandt in 1902 that all plant cells are totipotent and can be grown *in vitro* ultimately enabled the use of genetic engineering experimentally to knock out a gene or genes responsible for a specific trait, or to add genes such as GFP that report when a gene of interest is being expressed. These technologies enable the biotechnological use of whole plants or plant cell cultures grown in bioreactors to synthesise pesticides, antibiotics or other pharmaceuticals, as well as the practical application of genetically modified crops designed for traits such as improved yield.

Modern morphology recognises a continuum between the major morphological categories of root, stem (caulome), leaf (phyllome) and trichome. Furthermore, it emphasises structural dynamics. Modern systematics aims to reflect and discover phylogenetic relationships between plants. Modern Molecular phylogenetics largely ignores morphological characters, relying on DNA sequences as data. Molecular analysis of DNA sequences from most families of flowering plants enabled the Angiosperm Phylogeny Group to publish in 1998 a phylogeny of flowering plants, answering many of the questions about relationships among angiosperm families and species. The theoretical possibility of a practical method for identification of plant species and commercial varieties by DNA barcoding is the subject of active current research.

Scope and Importance

Botany involves the recording and description of plants, such as this herbarium specimen of the lady fern *Athyrium filix-femina*.

The study of plants is vital because they underpin almost all animal life on Earth by generating a large proportion of the oxygen and food that provide humans and other organisms with aerobic respiration with the chemical energy they need to exist. Plants, algae and cyanobacteria are the major groups of organisms that carry out photosynthesis, a process that uses the energy of sunlight to convert water and carbon dioxide into sugars that can be used both as a source of chemical energy and of organic molecules that are used in the structural components of cells. As a by-product of photosynthesis, plants release oxygen into the atmosphere, a gas that is required by nearly all living things to carry out cellular respiration. In addition, they are influential in the global carbon and water cycles and plant roots bind and stabilise soils, preventing soil erosion. Plants are crucial to the future of human society as they provide food, oxygen, medicine, and products for people, as well as creating and preserving soil.

Historically, all living things were classified as either animals or plants and botany covered the study of all organisms not considered animals. Botanists examine both the internal functions and processes within plant organelles, cells, tissues, whole plants, plant populations and plant communities. At each of these levels, a botanist may be concerned with the classification (taxonomy), phylogeny and evolution, structure (anatomy and morphology), or function (physiology) of plant life.

The strictest definition of "plant" includes only the "land plants" or embryophytes, which include seed plants (gymnosperms, including the pines, and flowering plants) and the free-sporing cryptogams including ferns, clubmosses, liverworts, hornworts and mosses. Embryophytes are multicellular eukaryotes descended from an ancestor that obtained its energy from sunlight by photosynthesis. They have life cycles with alternating haploid and diploid phases. The sexual haploid phase of embryophytes, known as the gametophyte, nurtures the developing diploid embryo sporophyte within its tissues for at least part of its life, even in the seed plants, where the gametophyte itself is nurtured by its parent sporophyte. Other groups of organisms that were previously studied by botanists include bacteria (now studied in bacteriology), fungi (mycology) – including lichen-forming fungi (lichenology), non-chlorophyte algae (phycology), and viruses (virology). However, attention is still given to these groups by botanists, and fungi (including lichens) and photosynthetic protists are usually covered in introductory botany courses.

Palaeobotanists study ancient plants in the fossil record to provide information about the evolutionary history of plants. Cyanobacteria, the first oxygen-releasing photosynthetic organisms on Earth, are thought to have given rise to the ancestor of plants by entering into an endosymbiotic relationship with an early eukaryote, ultimately becoming the chloroplasts in plant cells. The new photosynthetic plants (along with their algal relatives) accelerated the rise in atmospheric oxygen started by the cyanobacteria, changing the ancient oxygen-free, reducing, atmosphere to one in which free oxygen has been abundant for more than 2 billion years.

Among the important botanical questions of the 21st century are the role of plants as primary producers in the global cycling of life's basic ingredients: energy, carbon, oxygen, nitrogen and water, and ways that our plant stewardship can help address the global environmental issues of resource management, conservation, human food security, biologically invasive organisms, carbon sequestration, climate change, and sustainability.

Human Nutrition

Virtually all staple foods come either directly from primary production by plants, or indirectly from animals that eat them. Plants and other photosynthetic organisms are at the base of most food chains because they use the energy from the sun and nutrients from the soil and atmosphere, converting them into a form that can be used by animals. This is what ecologists call the first trophic level. The modern forms of the major staple foods, such as maize, rice, wheat and other cereal grasses, pulses, bananas and plantains, as well as flax and cotton grown for their fibres, are the outcome of prehistoric selection over thousands of years from among wild ancestral plants with the most desirable characteristics.

The food we eat comes directly or indirectly from plants such as rice.

Botanists study how plants produce food and how to increase yields, for example through plant breeding, making their work important to mankind's ability to feed the world and provide food security for future generations. Botanists also study weeds, which are a considerable problem in agriculture, and the biology and control of plant pathogens in agriculture and natural ecosystems. Ethnobotany is the study of the relationships between plants and people. When applied to the investigation of historical plant–people relationships ethnobotany may be referred to as archaeobotany or palaeoethnobotany. Some of the earliest plant-people relationships arose between the indigenous people of Canada in identifying edible plants from inedible plants. This relationship the indigenous people had with plants was recorded by ethnobotanists.

Plant Biochemistry

Plant biochemistry is the study of the chemical processes used by plants. Some of these processes are used in their primary metabolism like the photosynthetic Calvin cycle and crassulacean acid metabolism. Others make specialised materials like the cellulose and lignin used to build their bodies, and secondary products like resins and aroma compounds.

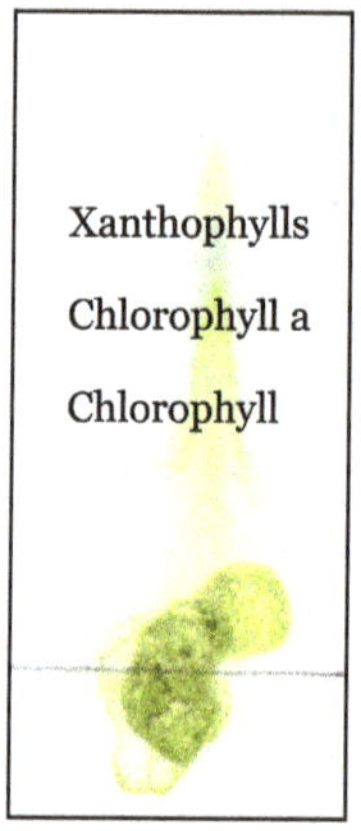

Plants make various photosynthetic pigments, some of which can be seen here through paper chromatography.

Plants and various other groups of photosynthetic eukaryotes collectively known as "algae" have unique organelles known as chloroplasts. Chloroplasts are thought to be descended from cyanobacteria that formed endosymbiotic relationships with ancient plant and algal ancestors. Chloroplasts and cyanobacteria contain the blue-green pigment chlorophyll *a*. Chlorophyll *a* (as well as its plant and green algal-specific cousin chlorophyll *b*) absorbs light in the blue-violet and orange/red parts of the spectrum while reflecting and transmitting the green light that we see as the characteristic colour of these organisms. The energy in the red and blue light that these pigments absorb is used by chloroplasts to make energy-rich carbon compounds from carbon dioxide and water by oxygenic photosynthesis, a process that generates molecular oxygen (O_2) as a by-product.

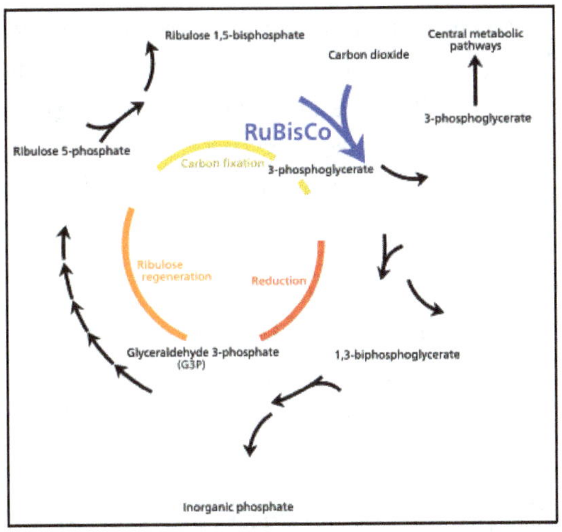

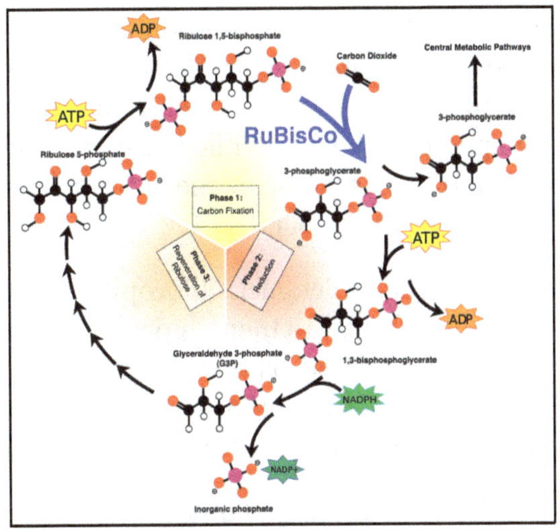

The Calvin cycle *(Interactive diagram)* The Calvin cycle incorporates carbon dioxide into sugar molecules.

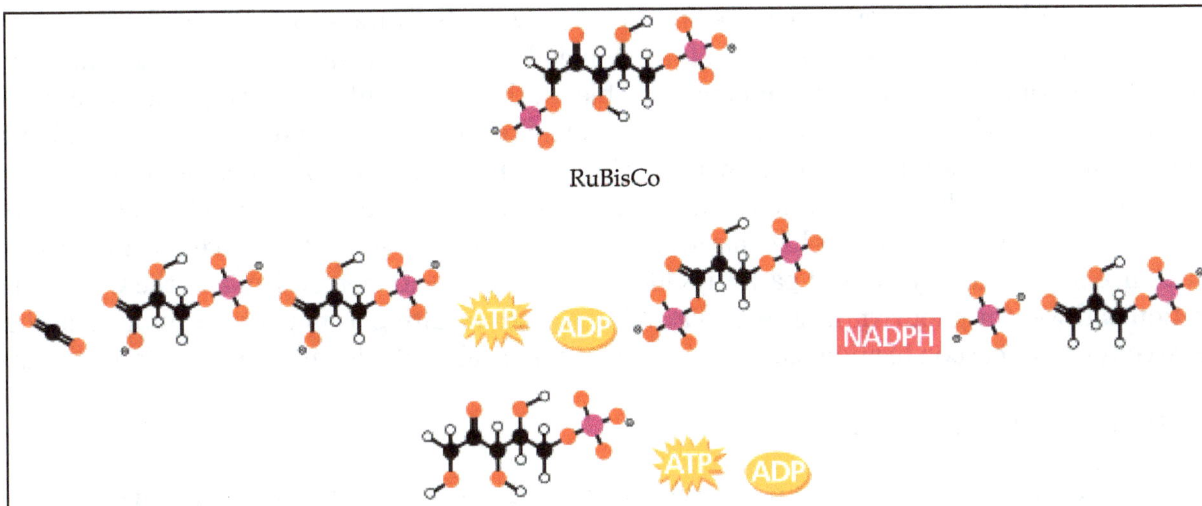

Carbon fixation
Reduction
3-phosphoglycerate
3-phosphoglycerate
Carbon dioxide
1,3-biphosphoglycerate
Glyceraldehyde-3-phosphate (G3P)
Inorganic phosphate
Ribulose 5-phosphate
Ribulose-1,5-bisphosphate
Edit · Source image

The light energy captured by chlorophyll *a* is initially in the form of electrons (and later a proton gradient) that's used to make molecules of ATP and NADPH which temporarily store and transport energy. Their energy is used in the light-independent reactions of the Calvin cycle by the enzyme rubisco to produce molecules of the 3-carbon sugar glyceraldehyde 3-phosphate (G3P). Glyceraldehyde 3-phosphate is the first product of photosynthesis and the raw material from which glucose and almost all other organic molecules of biological origin are synthesised. Some of the glucose is converted to starch which is stored in the chloroplast. Starch is the characteristic energy store of most land plants and algae, while inulin, a polymer of fructose is used for the same purpose in the sunflower family Asteraceae. Some of the glucose is converted to sucrose (common table sugar) for export to the rest of the plant.

Unlike in animals (which lack chloroplasts), plants and their eukaryote relatives have delegated many biochemical roles to their chloroplasts, including synthesising all their fatty acids, and most amino acids. The fatty acids that chloroplasts make are used for many things, such as providing material to build cell membranes out of and making the polymer cutin which is found in the plant cuticle that protects land plants from drying out.

Plants synthesise a number of unique polymers like the polysaccharide molecules cellulose, pectin and xyloglucan from which the land plant cell wall is constructed. Vascular land plants make lignin, a polymer used to strengthen the secondary cell walls of xylem tracheids and vessels to keep them from collapsing when a plant sucks water through them under water stress. Lignin is also used in other cell types like sclerenchyma fibres that provide structural support for a plant and is a major constituent of wood. Sporopollenin is a chemically resistant polymer found in the outer cell walls of spores and pollen of land plants responsible for the survival of early land plant spores and the pollen of seed plants in the fossil record. It is widely regarded as a marker for the start of land plant evolution during the Ordovician period. The concentration of carbon dioxide in the atmosphere today is much lower than it was when plants emerged onto land during the Ordovician and Silurian periods. Many monocots like maize and the pineapple and some dicots like the Asteraceae have since independently evolved pathways like Crassulacean acid metabolism and the C_4 carbon fixation pathway for photosynthesis which avoid the losses resulting from photorespiration in the more common C_3 carbon fixation pathway. These biochemical strategies are unique to land plants.

Medicine and Materials

Phytochemistry is a branch of plant biochemistry primarily concerned with the chemical substances produced by plants during secondary metabolism. Some of these compounds are toxins such as the alkaloid coniine from hemlock. Others, such as the essential oils peppermint oil and lemon oil are useful for their aroma, as flavourings and spices (e.g., capsaicin), and in medicine as pharmaceuticals as in opium from opium poppies. Many medicinal and recreational drugs, such as tetrahydrocannabinol (active ingredient in cannabis), caffeine, morphine and nicotine come directly from plants. Others are simple derivatives of botanical natural products. For example, the pain killer aspirin is the acetyl ester of salicylic acid, originally isolated from the bark of willow trees, and a wide range of opiate painkillers like heroin are obtained by chemical modification of morphine obtained from the opium poppy. Popular stimulants come from plants, such as caffeine from coffee, tea and chocolate, and nicotine from tobacco. Most alcoholic beverages come from fermentation of carbohydrate-rich plant products such as barley (beer), rice (sake) and grapes (wine).

Native Americans have used various plants as ways of treating illness or disease for thousands of years. This knowledge Native Americans have on plants has been recorded by enthnobotanists and then in turn has been used by pharmaceutical companies as a way of drug discovery.

Tapping a rubber tree in Thailand

Plants can synthesise useful coloured dyes and pigments such as the anthocyanins responsible for the red colour of red wine, yellow weld and blue woad used together to produce Lincoln green, indoxyl, source of the blue dye indigo traditionally used to dye denim and the artist's pigments gamboge and rose madder.

Sugar, starch, cotton, linen, hemp, some types of rope, wood and particle boards, papyrus and paper, vegetable oils, wax, and natural rubber are examples of commercially important materials made from plant tissues or their secondary products. Charcoal, a pure form of carbon made by pyrolysis of wood, has a long history as a metal-smelting fuel, as a filter material and adsorbent and as an artist's material and is one of the three ingredients of gunpowder. Cellulose, the world's most abundant organic polymer, can be converted into energy, fuels, materials and chemical feedstock. Products made from cellulose include rayon and cellophane, wallpaper paste, biobutanol and gun cotton. Sugarcane, rapeseed and soy are some of the plants with a highly fermentable sugar or oil content that are used as sources of biofuels, important alternatives to fossil fuels, such as biodiesel. Sweetgrass was used by NativeAmericanse to ward of bugs like mosquitoes. These bug repelling properties of sweetgrass were later found by the American Chemical Society in the molecules phytol and coumarin.

Plant Ecology

Holdridge life zones model relationships between vegetation type, moisture availability and temperature.

Plant ecology is the science of the functional relationships between plants and their habitats—the environments where they complete their life cycles. Plant ecologists study the composition of local and regional floras, their biodiversity, genetic diversity and fitness, the adaptation of plants to their environment, and their competitive or mutualistic interactions with other species. Some ecologists even rely on empirical data from indigenous people that is gathered by ethnobotanists. This information can relay a great deal of information on how the land once was thousands of years ago and how it has changed over that time. The goals of plant ecology are to understand the causes of their distribution patterns, productivity, environmental impact, evolution, and responses to environmental change.

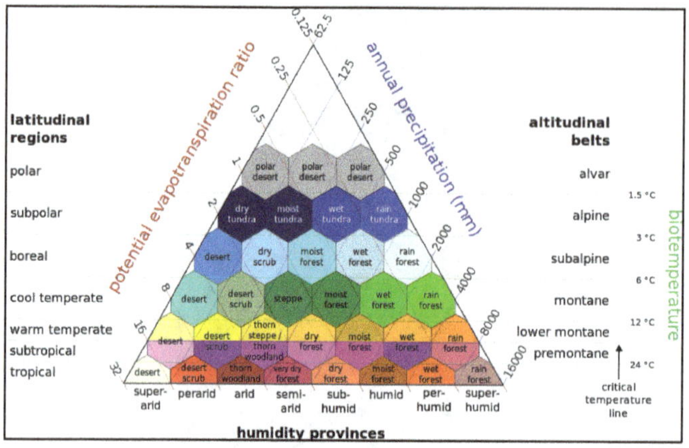

Plants depend on certain edaphic (soil) and climatic factors in their environment but can modify these factors too. For example, they can change their environment's albedo, increase runoff interception, stabilise mineral soils and develop their organic content, and affect local temperature. Plants compete with other organisms in their ecosystem for resources. They interact with their neighbours at a variety of spatial scales in groups, populations and communities that collectively constitute vegetation. Regions with characteristic vegetation types and dominant plants as well as similar abiotic and biotic factors, climate, and geography make up biomes like tundra or tropical rainforest.

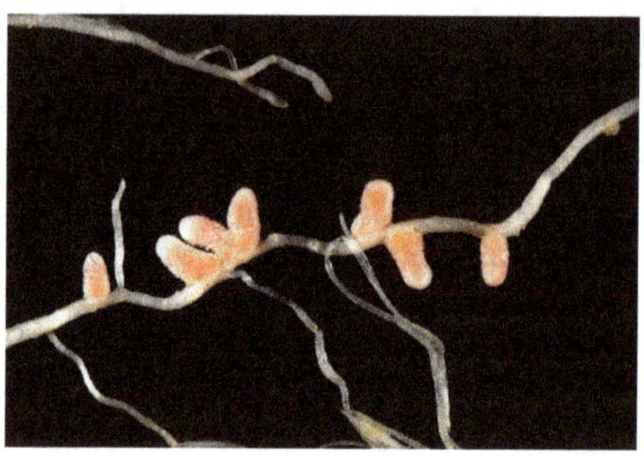

The nodules of *Medicago italica* contain the nitrogen fixing bacterium *Sinorhizobium meliloti*. The plant provides the bacteria with nutrients and an anaerobic environment, and the bacteria fix nitrogen for the plant.

Herbivores eat plants, but plants can defend themselves and some species are parasitic or even carnivorous. Other organisms form mutually beneficial relationships with plants. For example, mycorrhizal fungi and rhizobia provide plants with nutrients in exchange for food, ants are recruited by ant plants to provide protection, honey bees, bats and other animals pollinate flowers and humans and other animals act as dispersal vectors to spread spores and seeds.

Plants, Climate and Environmental Change

Plant responses to climate and other environmental changes can inform our understanding of how these changes affect ecosystem function and productivity. For example, plant phenology can be a useful proxy for temperature in historical climatology, and the biological impact of climate change and global warming. Palynology, the analysis of fossil pollen deposits in sediments from thousands or millions of years ago allows the reconstruction of past climates. Estimates of atmospheric CO_2 concentrations since the Palaeozoic have been obtained from stomatal densities and the leaf shapes and sizes of ancient land plants. Ozone depletion can expose plants to higher levels of ultraviolet radiation-B (UV-B), resulting in lower growth rates. Moreover, information from studies of community ecology, plant systematics, and taxonomy is essential to understanding vegetation change, habitat destruction and species extinction.

Genetics

Inheritance in plants follows the same fundamental principles of genetics as in other multicellular organisms. Gregor Mendel discovered the genetic laws of inheritance by studying inherited traits such as shape in *Pisum sativum* (peas). What Mendel learned from studying plants has had far reaching benefits outside of botany. Similarly, "jumping genes" were discovered by Barbara McClintock while she was studying maize. Nevertheless, there are some distinctive genetic differences between plants and other organisms.

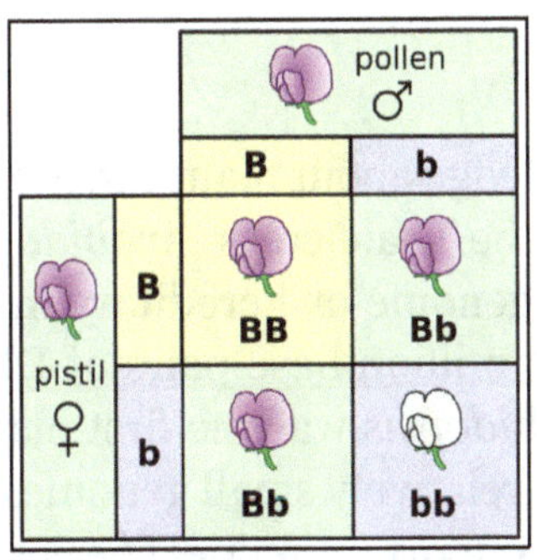

A Punnett square depicting a cross between two pea plants heterozygous for purple (B) and white (b) blossoms.

Species boundaries in plants may be weaker than in animals, and cross species hybrids are often possible. A familiar example is peppermint, *Mentha × piperita*, a sterile hybrid between *Mentha aquatica* and spearmint, *Mentha spicata*. The many cultivated varieties of wheat are the result of multiple inter- and intra-specific crosses between wild species and their hybrids. Angiosperms with monoecious flowers often have self-incompatibility mechanisms that operate between the pollen and stigma so that the pollen either fails to reach the stigma or fails to germinate and produce male gametes. This is one of several methods used by plants to promote outcrossing. In many land plants the male and female gametes are produced by separate individuals. These species are said to be dioecious when referring to vascular plant sporophytes and dioicous when referring to bryophyte gametophytes.

Unlike in higher animals, where parthenogenesis is rare, asexual reproduction may occur in plants by several different mechanisms. The formation of stem tubers in potato is one example. Particularly in arctic or alpine habitats, where opportunities for fertilisation of flowers by animals are rare, plantlets or bulbs, may develop instead of flowers, replacing sexual reproduction with asexual reproduction and giving rise to clonal populations genetically identical to the parent. This is one of several types of apomixis that occur in plants. Apomixis can also happen in a seed, producing a seed that contains an embryo genetically identical to the parent.

Most sexually reproducing organisms are diploid, with paired chromosomes, but doubling of their chromosome number may occur due to errors in cytokinesis. This can occur early in development to produce an autopolyploid or partly autopolyploid organism, or during normal processes of cellular differentiation to produce some cell types that are polyploid (endopolyploidy), or during gamete formation. An allopolyploid plant may result from a hybridisation event between two different species. Both autopolyploid and allopolyploid plants can often reproduce normally, but may be unable to cross-breed successfully with the parent population because there is a mismatch in chromosome numbers. These plants that are reproductively isolated from the parent species but live within the same geographical area, may be sufficiently successful to form a new species. Some otherwise sterile plant polyploids can still reproduce vegetatively or by seed apomixis, forming clonal populations of identical individuals. Durum wheat is a fertile tetraploid allopolyploid, while bread wheat is a fertile hexaploid. The commercial banana is an example of a sterile, seedless triploid hybrid. Common dandelion is a triploid that produces viable seeds by apomictic seed.

As in other eukaryotes, the inheritance of endosymbiotic organelles like mitochondria and chloroplasts in plants is non-Mendelian. Chloroplasts are inherited through the male parent in gymnosperms but often through the female parent in flowering plants.

Molecular Genetics

A considerable amount of new knowledge about plant function comes from studies of the molecular genetics of model plants such as the Thale cress, *Arabidopsis thaliana*, a weedy species in the mustard family (Brassicaceae). The genome or hereditary information contained in the genes of this species is encoded by about 135 million base pairs of DNA, forming one of the smallest genomes among flowering plants. *Arabidopsis* was the first plant to have its genome sequenced, in 2000. The sequencing of some other relatively small genomes, of rice (*Oryza sativa*) and *Brachypodium distachyon*, has made them important model species for understanding the genetics, cellular and molecular biology of cereals, grasses and monocots generally.

Thale cress, *Arabidopsis thaliana*, the first plant to have its genome sequenced, remains the most important model organism.

Model plants such as *Arabidopsis thaliana* are used for studying the molecular biology of plant cells and the chloroplast. Ideally, these organisms have small genomes that are well known or completely sequenced, small stature and short generation times. Corn has been used to study mechanisms of photosynthesis and phloem loading of sugar in C_4 plants. The single celled green alga *Chlamydomonas reinhardtii*, while not an embryophyte itself, contains a green-pigmented chloroplast related to that of land plants, making it useful for study. A red alga *Cyanidioschyzon merolae* has also been used to study some basic chloroplast functions. Spinach, peas, soybeans and a moss *Physcomitrella patens* are commonly used to study plant cell biology.

Agrobacterium tumefaciens, a soil rhizosphere bacterium, can attach to plant cells and infect them with a callus-inducing Ti plasmid by horizontal gene transfer, causing a callus infection called crown gall disease. Schell and Van Montagu (1977) hypothesised that the Ti plasmid could be a natural vector for introducing the Nif gene responsible for nitrogen fixation in the root nodules of legumes and other plant species. Today, genetic modification of the Ti plasmid is one of the main techniques for introduction of transgenes to plants and the creation of genetically modified crops.

Epigenetics

Epigenetics is the study of heritable changes in gene function that cannot be explained by changes in the underlying DNA sequence but cause the organism's genes to behave (or "express themselves") differently. One example of epigenetic change is the marking of the genes by DNA methylation which determines whether they will be expressed or not. Gene expression can also be controlled by repressor proteins that attach to silencer regions of the DNA and prevent that region of the DNA code from being expressed. Epigenetic marks may be added or removed from the DNA during programmed stages of development of the plant, and are responsible, for example, for the differences between anthers, petals and normal leaves, despite the fact that they all have the same underlying genetic code. Epigenetic changes may be temporary or may remain through successive cell divisions for the remainder of the cell's life. Some epigenetic changes have been shown to be heritable, while others are reset in the germ cells.

Epigenetic changes in eukaryotic biology serve to regulate the process of cellular differentiation. During morphogenesis, totipotent stem cells become the various pluripotent cell lines of the embryo, which in turn become fully differentiated cells. A single fertilised egg cell, the zygote, gives rise to the many different plant cell types including parenchyma, xylem vessel elements, phloem sieve tubes, guard cells of the epidermis, etc. as it continues to divide. The process results from the epigenetic activation of some genes and inhibition of others.

Unlike animals, many plant cells, particularly those of the parenchyma, do not terminally differentiate, remaining totipotent with the ability to give rise to a new individual plant. Exceptions include highly lignified cells, the sclerenchyma and xylem which are dead at maturity, and the phloem sieve tubes which lack nuclei. While plants use many of the same epigenetic mechanisms as animals, such as chromatin remodelling, an alternative hypothesis is that plants set their gene expression patterns using positional information from the environment and surrounding cells to determine their developmental fate.

Plant Evolution

The chloroplasts of plants have a number of biochemical, structural and genetic similarities to cyanobacteria, (commonly but incorrectly known as "blue-green algae") and are thought to be derived from an ancient endosymbiotic relationship between an ancestral eukaryotic cell and a cyanobacterial resident.

The algae are a polyphyletic group and are placed in various divisions, some more closely related to plants than others. There are many differences between them in features such as cell wall composition, biochemistry, pigmentation, chloroplast structure and nutrient reserves. The algal division Charophyta, sister to the green algal division Chlorophyta, is considered to contain the ancestor of true plants. The Charophyte class Charophyceae and the land plant sub-kingdom Embryophyta together form the monophyletic group or clade Streptophytina.

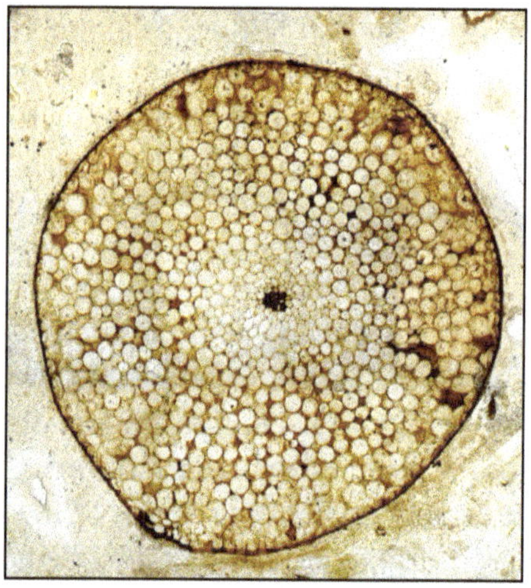

Transverse section of a fossil stem of the Devonian vascular plant *Rhynia gwynne-vaughani*.

Nonvascular land plants are embryophytes that lack the vascular tissues xylem and phloem. They include mosses, liverworts and hornworts. Pteridophytic vascular plants with true xylem and phloem that reproduced by spores germinating into free-living gametophytes evolved during the Silurian period and diversified into several lineages during the late Silurian and early Devonian. Representatives of the lycopods have survived to the present day. By the end of the Devonian period, several groups, including the lycopods, sphenophylls and progymnosperms, had independently evolved "megaspory" – their spores were of two distinct sizes, larger megaspores and smaller microspores. Their reduced gametophytes developed from megaspores retained within the spore-producing organs (megasporangia) of the sporophyte, a condition known as endospory. Seeds consist of an endosporic megasporangium surrounded by one or two sheathing layers (integuments). The young sporophyte develops within the seed, which on germination splits to release it. The earliest known seed plants date from the latest Devonian Famennian stage. Following the evolution of the seed habit, seed plants diversified, giving rise to a number of now-extinct groups, including seed ferns, as well as the modern gymnosperms and angiosperms. Gymnosperms produce "naked seeds" not fully enclosed in an ovary; modern representatives include conifers, cycads, *Ginkgo*, and Gnetales. Angiosperms produce seeds enclosed in a structure such as a carpel or an ovary. Ongoing research on the molecular phylogenetics of living plants appears to show that the angiosperms are a sister clade to the gymnosperms.

Plant Physiology

Plant physiology encompasses all the internal chemical and physical activities of plants associated with life. Chemicals obtained from the air, soil and water form the basis of all plant metabolism. The energy of sunlight, captured by oxygenic photosynthesis and released by cellular respiration, is the basis of almost all life. Photoautotrophs, including all green plants, algae and cyanobacteria gather energy directly from sunlight by photosynthesis. Heterotrophs including all animals, all fungi, all completely parasitic plants, and non-photosynthetic bacteria take in organic molecules produced by photoautotrophs and respire them or use them in the construction of cells and tissues. Respiration is the oxidation of carbon compounds by breaking them down into simpler structures to release the energy they contain, essentially the opposite of photosynthesis.

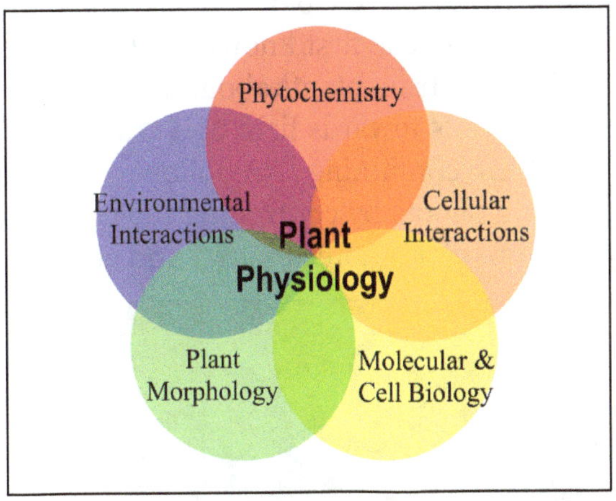

Five of the key areas of study within plant physiology.

Molecules are moved within plants by transport processes that operate at a variety of spatial scales. Subcellular transport of ions, electrons and molecules such as water and enzymes occurs across cell membranes. Minerals and water are transported from roots to other parts of the plant in the transpiration stream. Diffusion, osmosis, and active transport and mass flow are all different ways transport can occur. Examples of elements that plants need to transport are nitrogen, phosphorus, potassium, calcium, magnesium, and sulphur. In vascular plants, these elements are extracted from the soil as soluble ions by the roots and transported throughout the plant in the xylem. Most of the elements required for plant nutrition come from the chemical breakdown of soil minerals. Sucrose produced by photosynthesis is transported from the leaves to other parts of the plant in the phloem and plant hormones are transported by a variety of processes.

Plant Hormones

Plants are not passive, but respond to external signals such as light, touch, and injury by moving or growing towards or away from the stimulus, as appropriate. Tangible evidence of touch sensitivity is the almost instantaneous collapse of leaflets of *Mimosa pudica*, the insect traps of Venus flytrap and bladderworts, and the pollinia of orchids.

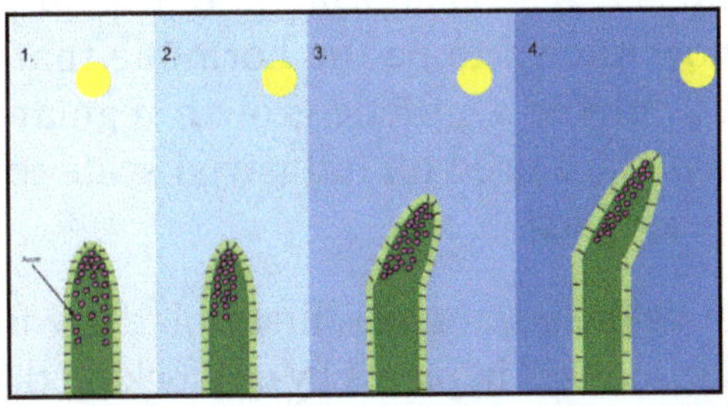

1 An oat coleoptile with the sun overhead. Auxin (pink) is evenly distributed in its tip.
2 With the sun at an angle and only shining on one side of the shoot, auxin moves to the opposite side and stimulates cell elongation there.
3 and **4** Extra growth on that side causes the shoot to bend towards the sun.

The hypothesis that plant growth and development is coordinated by plant hormones or plant growth regulators first emerged in the late 19th century. Darwin experimented on the movements of plant shoots and roots towards light and gravity, and concluded "It is hardly an exaggeration to say that the tip of the radicle . . acts like the brain of one of the lower animals . . directing the several movements". About the same time, the role of auxins in control of plant growth was first outlined by the Dutch scientist Frits Went. The first known auxin, indole-3-acetic acid (IAA), which promotes cell growth, was only isolated from plants about 50 years later. This compound mediates the tropic responses of shoots and roots towards light and gravity. The finding in 1939 that plant callus could be maintained in culture containing IAA, followed by the observation in 1947 that it could be induced to form roots and shoots by controlling the concentration of growth hormones were key steps in the development of plant biotechnology and genetic modification.

Venus's fly trap, *Dionaea muscipula,* showing the touch-sensitive insect trap in action.

Cytokinins are a class of plant hormones named for their control of cell division or cytokinesis. The natural cytokinin zeatin was discovered in corn, *Zea mays*, and is a derivative of the purine adenine. Zeatin is produced in roots and transported to shoots in the xylem where it promotes cell division, bud development, and the greening of chloroplasts. The gibberelins, such as Gibberelic acid are diterpenes synthesised from acetyl CoA via the mevalonate pathway. They are involved in the promotion of germination and dormancy-breaking in seeds, in regulation of plant height by controlling stem elongation and the control of flowering. Abscisic acid (ABA) occurs in all land plants except liverworts, and is synthesised from carotenoids in the chloroplasts and other plastids. It inhibits cell division, promotes seed maturation, and dormancy, and promotes stomatal closure. It was so named because it was originally thought to control abscission. Ethylene is a gaseous hormone that is produced in all higher plant tissues from methionine. It is now known to be the hormone that stimulates or regulates fruit ripening and abscission, and it, or the synthetic growth regulator ethephon which is rapidly metabolised to produce ethylene, are used on industrial scale to promote ripening of cotton, pineapples and other climacteric crops.

Another class of phytohormones is the jasmonates, first isolated from the oil of *Jasminum grandiflorum* which regulates wound responses in plants by unblocking the expression of genes required in the systemic acquired resistance response to pathogen attack.

In addition to being the primary energy source for plants, light functions as a signalling device, providing information to the plant, such as how much sunlight the plant receives each day. This can result in adaptive changes in a process known as photomorphogenesis. Phytochromes are the photoreceptors in a plant that are sensitive to light.

Plant Anatomy and Morphology

Plant anatomy is the study of the structure of plant cells and tissues, whereas plant morphology is the study of their external form. All plants are multicellular eukaryotes, their DNA stored in nuclei. The characteristic features of plant cells that distinguish them from those of animals and fungi include a primary cell wall composed of the polysaccharides cellulose, hemicellulose and pectin, larger vacuoles than in animal cells and the presence of plastids with unique photosynthetic and biosynthetic functions as in the chloroplasts. Other plastids contain storage products such as starch (amyloplasts) or lipids (elaioplasts). Uniquely, streptophyte cells and those of the green algal order Trentepohliales divide by construction of a phragmoplast as a template for building a cell plate late in cell division.

A nineteenth-century illustration showing the morphology of the roots, stems, leaves and flowers of the rice plant *Oryza sativa*.

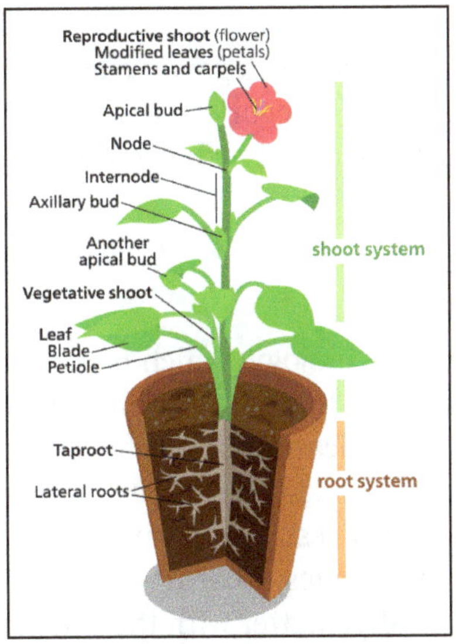

A diagram of a "typical" eudicot, the most common type of plant (three-fifths of all plant species). No plant actually looks exactly like this though.

The bodies of vascular plants including clubmosses, ferns and seed plants (gymnosperms and angiosperms) generally have aerial and subterranean subsystems. The shoots consist of stems bearing green photosynthesising leaves and reproductive structures. The underground vascularised roots bear root hairs at their tips and generally lack chlorophyll. Non-vascular plants, the liverworts, hornworts and mosses do not produce ground-penetrating vascular roots and most of the plant participates in photosynthesis. The sporophyte generation is nonphotosynthetic in liverworts but may be able to contribute part of its energy needs by photosynthesis in mosses and hornworts.

The root system and the shoot system are interdependent – the usually nonphotosynthetic root system depends on the shoot system for food, and the usually photosynthetic shoot system depends on water and minerals from the root system. Cells in each system are capable of creating cells of the other and producing adventitious shoots or roots. Stolons and tubers are examples of shoots that can grow roots. Roots that spread out close to the surface, such as those of willows, can produce shoots and ultimately new plants. In the event that one of the systems is lost, the other can often regrow it. In fact it is possible to grow an entire plant from a single leaf, as is the case with *Saintpaulia*, or even a single cell – which can dedifferentiate into a callus (a mass of unspecialised cells) that can grow into a new plant. In vascular plants, the xylem and phloem are the conductive tissues that transport resources between shoots and roots. Roots are often adapted to store food such as sugars or starch, as in sugar beets and carrots.

Stems mainly provide support to the leaves and reproductive structures, but can store water in succulent plants such as cacti, food as in potato tubers, or reproduce vegetatively as in the stolons of strawberry plants or in the process of layering. Leaves gather sunlight and carry out photosynthesis. Large, flat, flexible, green leaves are called foliage leaves. Gymnosperms, such as conifers, cycads, *Ginkgo*, and gnetophytes are seed-producing plants with open seeds. Angiosperms are seed-producing plants that produce flowers and have enclosed seeds. Woody plants, such as azaleas and oaks, undergo a secondary growth phase resulting in two additional types of tissues: wood (secondary xylem) and bark (secondary phloem and cork). All gymnosperms and many angiosperms are woody plants. Some plants reproduce sexually, some asexually, and some via both means.

Although reference to major morphological categories such as root, stem, leaf, and trichome are useful, one has to keep in mind that these categories are linked through intermediate forms so that a continuum between the categories results. Furthermore, structures can be seen as processes, that is, process combinations.

Systematic Botany

Systematic botany is part of systematic biology, which is concerned with the range and diversity of organisms and their relationships, particularly as determined by their evolutionary history. It involves, or is related to, biological classification, scientific taxonomy and phylogenetics. Biological classification is the method by which botanists group organisms into categories such as genera or species. Biological classification is a form of scientific taxonomy. Modern taxonomy is rooted in the work of Carl Linnaeus, who grouped species according to shared physical characteristics. These groupings have since been revised to align better with the Darwinian principle of common descent – grouping organisms by ancestry rather than superficial characteristics. While scientists do not always agree on how to classify organisms, molecular phylogenetics, which uses DNA sequences as data, has driven many recent revisions along evolutionary lines and is likely to continue to do so. The dominant classification system is called Linnaean taxonomy. It includes ranks and binomial nomenclature. The nomenclature of botanical organisms is codified in the International Code of Nomenclature for algae, fungi, and plants (ICN) and administered by the International Botanical Congress.

A botanist preparing a plant specimen for mounting in the herbarium.

Kingdom Plantae belongs to Domain Eukarya and is broken down recursively until each species is separately classified. The order is: Kingdom; Phylum (or Division); Class; Order; Family; Genus (plural *genera*); Species. The scientific name of a plant represents its genus and its species within the genus, resulting in a single worldwide name for each organism. For example, the tiger lily is *Lilium columbianum*. *Lilium* is the genus, and *columbianum* the specific epithet. The combination is the name of the species. When writing the scientific name of an organism, it is proper to capitalise the first letter in the genus and put all of the specific epithet in lowercase. Additionally, the entire term is ordinarily italicised (or underlined when italics are not available).

The evolutionary relationships and heredity of a group of organisms is called its phylogeny. Phylogenetic studies attempt to discover phylogenies. The basic approach is to use similarities based on shared inheritance to determine relationships. As an example, species of *Pereskia* are trees or bushes with prominent leaves. They do not obviously resemble a typical leafless cactus such as an *Echinocactus*. However, both *Pereskia* and *Echinocactus* have spines produced from areoles (highly specialised pad-like structures) suggesting that the two genera are indeed related.

Two Cacti of Very Different Appearance

Judging relationships based on shared characters requires care, since plants may resemble one another through convergent evolution in which characters have arisen independently. Some euphorbias have leafless, rounded bodies adapted to water conservation similar to those of globular cacti, but characters such as the structure of their flowers make it clear that the two groups are not closely related. The cladistic method takes a systematic approach to characters, distinguishing between those that carry no information about shared evolutionary history – such as those evolved separately in different groups (homoplasies) or those left over from ancestors (plesiomorphies) – and derived characters, which have been passed down from innovations in a shared ancestor (apomorphies). Only derived characters, such as the spine-producing areoles of cacti, provide evidence for descent from a common ancestor. The results of cladistic analyses are expressed as cladograms: tree-like diagrams showing the pattern of evolutionary branching and descent.

Echinocactus grusonii

Pereskia aculeata

Although *Pereskia* is a tree with leaves, it has spines and areoles like a more typical cactus, such as *Echinocactus*.

From the 1990s onwards, the predominant approach to constructing phylogenies for living plants has been molecular phylogenetics, which uses molecular characters, particularly DNA sequences, rather than morphological characters like the presence or absence of spines and areoles. The difference is that the genetic code itself is used to decide evolutionary relationships, instead of being used indirectly via the characters it gives rise to. Clive Stace describes this as having "direct access to the genetic basis of evolution." As a simple example, prior to the use of genetic evidence, fungi were thought either to be plants or to be more closely related to plants than animals. Genetic evidence suggests that the true evolutionary relationship of multicelled organisms is as shown in the cladogram below – fungi are more closely related to animals than to plants.

In 1998 the Angiosperm Phylogeny Group published a phylogeny for flowering plants based on an analysis of DNA sequences from most families of flowering plants. As a result of this work, many questions, such as which families represent the earliest branches of angiosperms, have now been answered. Investigating how plant species are related to each other allows botanists to better understand the process of evolution in plants. Despite the study of model plants and increasing use of DNA evidence, there is ongoing work and discussion among taxonomists about how best to classify plants into various taxa. Technological developments such as computers and electron microscopes have greatly increased the level of detail studied and speed at which data can be analysed.

Plant Anatomy

Plant anatomy or phytotomy is the general term for the study of the internal structure of plants. While originally it included plant morphology, which is the description of the physical form and external structure of plants, since the mid-20th century the investigations of plant anatomy are considered a separate, distinct field, and plant anatomy refers to *just* the internal plant structures. Plant anatomy is now frequently investigated at the cellular level, and often involves the sectioning of tissues and microscopy.

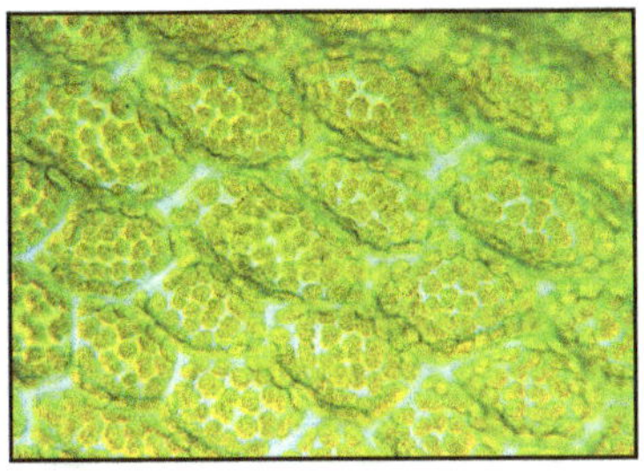

Chloroplasts in leaf cells of the moss *Mnium stellare.*

Structural Divisions

Plant anatomy is often divided into the following categories:

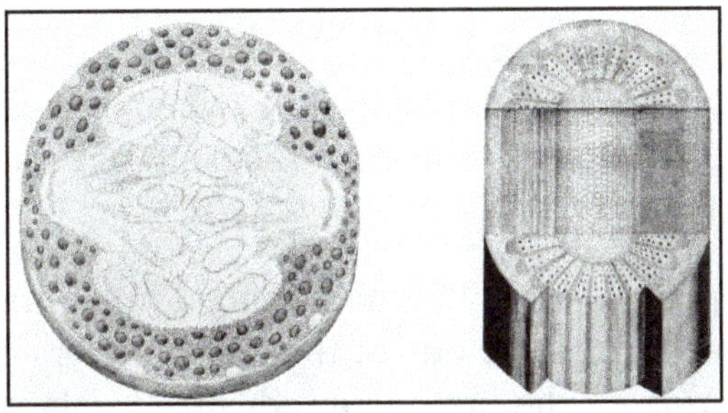

Vascular tissue of a gooseberry vine from Grew's *Anatomy of Plants.*

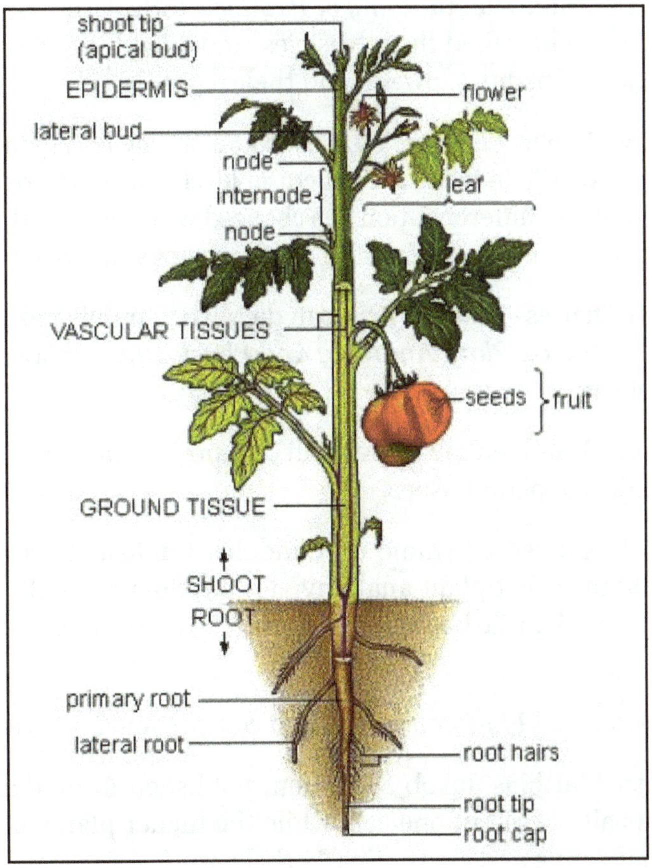

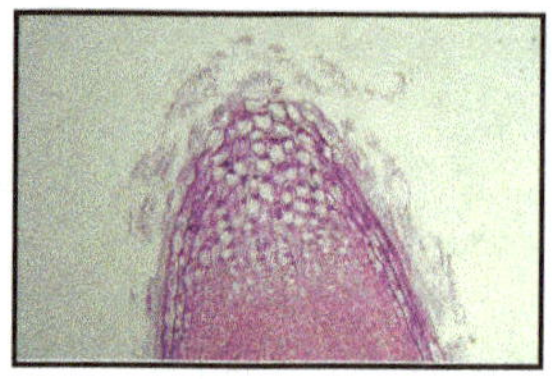

Root tip

History

About 300 BC Theophrastus wrote a number of plant treatises, only two of which survive, *Enquiry into Plants* (Περ⍰ φυτ⍰ν ⍰στορία), and *On the Causes of Plants* (Περ⍰ φυτ⍰ν α⍰τι⍰ν). He developed concepts of plant morphology and classification, which did not withstand the scientific scrutiny of the Renaissance.

A Swiss physician and botanist, Gaspard Bauhin, introduced binomial nomenclature into plant taxonomy. He published *Pinax theatri botanici* in 1596, which was the first to use this convention for naming of species. His criteria for classification included natural relationships, or 'affinities', which in many cases were structural.

It was in the late 1600s that plant anatomy became refined into a modern science. Italian doctor and microscopist, Marcello Malpighi, was one of the two founders of plant anatomy. In 1671 he published his *Anatomia Plantarum*, the first major advance in plant physiogamy since Aristotle. The other founder was the British doctor Nehemiah Grew. He published *An Idea of a Philosophical History of Plants* in 1672 and *The Anatomy of Plants* in 1682. Grew is credited with the recognition of plant cells, although he called them 'vesicles' and 'bladders'. He correctly identified and described the sexual organs of plants (flowers) and their parts.

In the Eighteenth Century, Carolus Linnaeus established taxonomy based on structure, and his early work was with plant anatomy. While the exact structural level which is to be considered to be scientifically valid for comparison and differentiation has changed with the growth of knowledge, the basic principles were established by Linnaeus. He published his master work, *Species Plantarum* in 1753.

In 1802, French botanist Charles-François Brisseau de Mirbel, published *Traité d'anatomie et de physiologie végétale* (*Treatise on Plant Anatomy and Physiology*) establishing the beginnings of the science of plant cytology.

In 1812, Johann Jacob Paul Moldenhawer published *Beyträge zur Anatomie der Pflanzen*, describing microscopic studies of plant tissues.

In 1813 a Swiss botanist, Augustin Pyrame de Candolle, published *Théorie élémentaire de la botanique*, in which he argued that plant anatomy, not physiology, ought to be the sole basis for plant classification. Using a scientific basis, he established structural criteria for defining and separating plant genera.

In 1830, Franz Meyen published *Phytotomie*, the first comprehensive review of plant anatomy.

In 1838 German botanist Matthias Jakob Schleiden, published *Contributions to Phytogenesis*, stating, "the lower plants all consist of one cell, while the higher plants are composed of (many) individual cells" thus confirming and continuing Mirbel's work.

A German-Polish botanist, Eduard Strasburger, described the mitotic process in plant cells and further demonstrated that new cell nuclei can only arise from the division of other pre-existing nuclei. His *Studien über Protoplasma* was published in 1876.

Gottlieb Haberlandt, a German botanist, studied plant physiology and classified plant tissue based upon function. On this basis, in 1884 he published *Physiologische Pflanzenanatomie* (*Physiological Plant Anatomy*) in which he described twelve types of tissue systems (absorptive, mechanical, photosynthetic, etc.).

British paleobotanists Dunkinfield Henry Scott and William Crawford Williamson described the structures of fossilized plants at the end of the Nineteenth Century. Scott's *Studies in Fossil Botany* was published in 1900.

Following Charles Darwin's *Origin of Species* a Canadian botanist, Edward Charles Jeffrey, who was studying the comparative anatomy and phylogeny of different vascular plant groups, applied the theory to plants using the form and structure of plants to establish a number of evolutionary lines. He published his *The Anatomy of Woody Plants* in 1917.

The growth of comparative plant anatomy was spearheaded by British botanist Agnes Arber. She published *Water Plants: A Study of Aquatic Angiosperms* in 1920, *Monocotyledons: A Morphological Study* in 1925, and *The Gramineae: A Study of Cereal, Bamboo and Grass* in 1934.

Following World War II, Katherine Esau published, *Plant Anatomy* (1953), which became the definitive textbook on plant structure in North American universities and elsewhere, it was still in print as of 2006. She followed up with her *Anatomy of seed plants* in 1960.

Plant Breeding

Plant breeding is the art and science of changing the traits of plants in order to produce desired characteristics. Plant breeding can be accomplished through many different techniques ranging from simply selecting plants with desirable characteristics for propagation, to more complex mo-lecular techniques.

The Yecoro wheat (right) cultivar is sensitive to salinity, plants resulting from a hybrid cross with cultivar W4910 (left) show greater tolerance to high salinity.

Plant breeding has been practiced for thousands of years, since near the beginning of human civilization. It is practiced worldwide by individuals such as gardeners and farmers, or by professional plant breeders employed by organizations such as government institutions, universities, crop-specific industry associations or research centers.

International development nation agencies believe that breeding new crops is important for ensuring food security by developing new varieties that are higher-yielding, disease resistant, drought-resistant or regionally adapted to different environments and growing conditions.

History

Plant breeding started with sedentary agriculture and particularly the domestication of the first agricultural plants, a practice which is estimated to date back 9,000 to 11,000 years. Initially early farmers simply selected food plants with particular desirable characteristics, and employed these as progenitors for subsequent generations, resulting in an accumulation of valuable traits over time.

Gregor Mendel's experiments with plant hybridization led to his establishing laws of inheritance. Once this work became well known, it formed the basis of the new science of genetics, which stimulated research by many plant scientists dedicated to improving crop production through plant breeding.

Modern plant breeding is applied genetics, but its scientific basis is broader, covering molecular biology, cytology, systematics, physiology, pathology, entomology, chemistry, and statistics (biometrics). It has also developed its own technology.

Classical Plant Breeding

One major technique of plant breeding is selection, the process of selectively propagating plants with desirable characteristics and eliminating or "culling" those with less desirable characteristics.

Another technique is the deliberate interbreeding (crossing) of closely or distantly related individuals to produce new crop varieties or lines with desirable properties. Plants are crossbred to introduce traits/genes from one variety or line into a new genetic background. For example, a mildew-resistant pea may be crossed with a high-yielding but susceptible pea, the goal of the cross being to introduce mildew resistance without losing the high-yield characteristics. Progeny from the cross would then be crossed with the high-yielding parent to ensure that the progeny were most like the high-yielding parent, (backcrossing). The progeny from that cross would then be tested for yield (selection, as described above) and mildew resistance and high-yielding resistant plants would be further developed. Plants may also be crossed with themselves to produce inbred varieties for breeding. Pollinators may be excluded through the use of pollination bags.

Classical breeding relies largely on homologous recombination between chromosomes to generate genetic diversity. The classical plant breeder may also make use of a number of *in vitro* techniques such as protoplast fusion, embryo rescue or mutagenesis (see below) to generate diversity and produce hybrid plants that would not exist in nature.

Traits that breeders have tried to incorporate into crop plants include:

1. Improved quality, such as increased nutrition, improved flavor, or greater beauty.

2. Increased yield of the crop.
3. Increased tolerance of environmental pressures (salinity, extreme temperature, drought).
4. Resistance to viruses, fungi and bacteria.
5. Increased tolerance to insect pests.
6. Increased tolerance of herbicides.
7. Longer storage period for the harvested crop.

Before World War II

Successful commercial plant breeding concerns were founded from the late 19th century. Gartons Agricultural Plant Breeders in England was established in the 1890s by John Garton, who was one of the first to commercialize new varieties of agricultural crops created through cross-pollination. The firm's first introduction was Abundance Oat, one of the first agricultural grain varieties bred from a *controlled* cross, introduced to commerce in 1892.

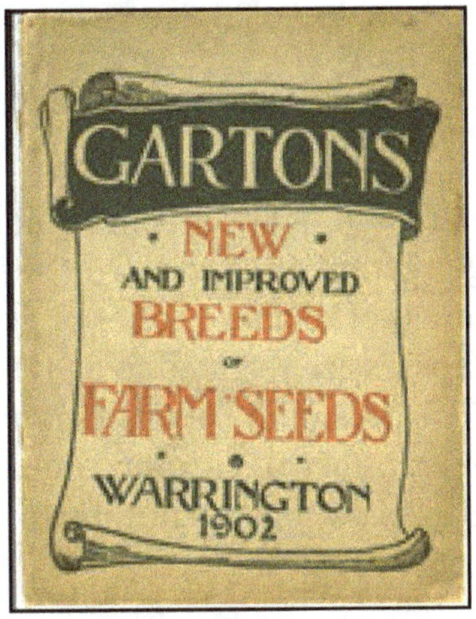

Garton's catalogue from 1902.

In the early 20th century, plant breeders realized that Mendel's findings on the non-random nature of inheritance could be applied to seedling populations produced through deliberate pollinations to predict the frequencies of different types. Wheat hybrids were bred to increase the crop production of Italy during the so-called "Battle for Grain" (1925–1940). Heterosis was explained by George Harrison Shull. It describes the tendency of the progeny of a specific cross to outperform both parents. The detection of the usefulness of heterosis for plant breeding has led to the development of inbred lines that reveal a heterotic yield advantage when they are crossed. Maize was the first species where heterosis was widely used to produce hybrids.

Statistical methods were also developed to analyze gene action and distinguish heritable variation from variation caused by environment. In 1933 another important breeding technique, cytoplasmic male sterility (CMS), developed in maize, was described by Marcus Morton Rhoades. CMS is a maternally inherited trait that makes the plant produce sterile pollen. This enables the production of hybrids without the need for labor-intensive detasseling.

These early breeding techniques resulted in large yield increase in the United States in the early 20th century. Similar yield increases were not produced elsewhere until after World War II, the Green Revolution increased crop production in the developing world in the 1960s.

After World War II

Following World War II a number of techniques were developed that allowed plant breeders to hybridize distantly related species, and artificially induce genetic diversity.

In vitro-culture of Vitis (grapevine), Geisenheim Grape Breeding Institute.

When distantly related species are crossed, plant breeders make use of a number of plant tissue culture techniques to produce progeny from otherwise fruitless mating. Interspecific and intergeneric hybrids are produced from a cross of related species or genera that do not normally sexually reproduce with each other. These crosses are referred to as *Wide crosses*. For example, the cereal triticale is a wheat and rye hybrid. The cells in the plants derived from the first generation created from the cross contained an uneven number of chromosomes and as result was sterile. The cell division inhibitor colchicine was used to double the number of chromosomes in the cell and thus allow the production of a fertile line.

Failure to produce a hybrid may be due to pre- or post-fertilization incompatibility. If fertilization is possible between two species or genera, the hybrid embryo may abort before maturation. If this does occur the embryo resulting from an interspecific or intergeneric cross can sometimes be rescued and cultured to produce a whole plant. Such a method is referred to as Embryo Rescue. This technique has been used to produce new rice for Africa, an interspecific cross of Asian rice *(Oryza sativa)* and African rice *(Oryza glaberrima)*.

Hybrids may also be produced by a technique called protoplast fusion. In this case protoplasts are fused, usually in an electric field. Viable recombinants can be regenerated in culture.

Chemical mutagens like EMS and DMS, radiation and transposons are used to generate mutants with desirable traits to be bred with other cultivars - a process known as *Mutation Breeding*. Classical plant breeders also generate genetic diversity within a species by exploiting a process called somaclonal variation, which occurs in plants produced from tissue culture, particularly plants derived from callus. Induced polyploidy, and the addition or removal of chromosomes using a technique called chromosome engineering may also be used.

When a desirable trait has been bred into a species, a number of crosses to the favored parent are made to make the new plant as similar to the favored parent as possible. Returning to the example of the mildew resistant pea being crossed with a high-yielding but susceptible pea, to make the mildew resistant progeny of the cross most like the high-yielding parent, the progeny will be crossed back to that parent for several generations. This process removes most of the genetic contribution of the mildew resistant parent. Classical breeding is therefore a cyclical process.

With classical breeding techniques, the breeder does not know exactly what genes have been introduced to the new cultivars. Some scientists therefore argue that plants produced by classical breeding methods should undergo the same safety testing regime as genetically modified plants. There have been instances where plants bred using classical techniques have been unsuitable for human consumption, for example the poison solanine was unintentionally increased to unacceptable levels in certain varieties of potato through plant breeding. New potato varieties are often screened for solanine levels before reaching the marketplace.

Modern Plant Breeding

Modern plant breeding may use techniques of molecular biology to select, or in the case of genetic modification, to insert, desirable traits into plants. Application of biotechnology or molecular biol-ogy is also known as molecular breeding.

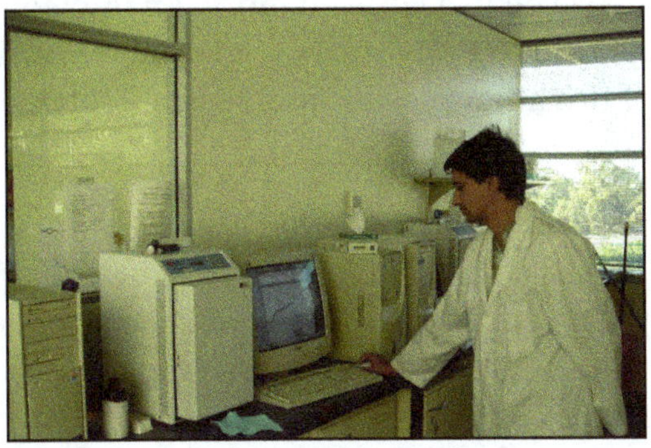

Modern facilities in molecular biology have converted classical plant breeding to molecular plant breeding.

Steps of Plant Breeding

The following are the major activities of plant breeding:

1. Collection of variation,
2. Selection,
3. Evaluation,
4. Release,
5. Multiplication,
6. Distribution of the new variety,
7. Selling to people.

Marker Assisted Selection

Sometimes many different genes can influence a desirable trait in plant breeding. The use of tools such as molecular markers or DNA fingerprinting can map thousands of genes. This allows plant breeders to screen large populations of plants for those that possess the trait of interest. The screening is based on the presence or absence of a certain gene as determined by laboratory procedures, rather than on the visual identification of the expressed trait in the plant.

Reverse Breeding and Doubled Haploids (DH)

A method for efficiently producing homozygous plants from a heterozygous starting plant, which has all desirable traits. This starting plant is induced to produce doubled haploid from haploid cells, and later on creating homozygous/doubled haploid plants from those cells. While in natural offspring genetic recombination occurs and traits can be unlinked from each other, in doubled haploid cells and in the resulting DH plants recombination is no longer an issue. There, a recombination between two corresponding chromosomes does not lead to un-linkage of alleles or traits, since it just leads to recombination with its identical copy. Thus, traits on one chromosome stay linked. Selecting those offspring having the desired set of chromosomes and crossing them will result in a final F1 hybrid plant, having exactly the same set of chromosomes, genes and traits as the starting hybrid plant. The homozygous parental lines can reconstitute the original heterozygous plant by crossing, if desired even in a large quantity. An individual heterozygous plant can be converted into a heterozygous variety (F1 hybrid) without the necessity of vegetative propagation but as the result of the cross of two homozygous/doubled haploid lines derived from the originally selected plant. patent

Genetic Modification

Genetic modification of plants is achieved by adding a specific gene or genes to a plant, or by knocking down a gene with RNAi, to produce a desirable phenotype. The plants resulting from adding a gene are often referred to as transgenic plants. If for genetic modification genes of the species or of a crossable plant are used under control of their native promoter, then they are called cisgenic plants. Sometimes genetic modification can produce a plant with the desired trait or traits faster than classical breeding because the majority of the plant's genome is not altered.

To genetically modify a plant, a genetic construct must be designed so that the gene to be added or removed will be expressed by the plant. To do this, a promoter to drive transcription and a termination sequence to stop transcription of the new gene, and the gene or genes of interest must be introduced to the plant. A marker for the selection of transformed plants is also included. In the laboratory, antibiotic resistance is a commonly used marker: Plants that have been successfully transformed will grow on media containing antibiotics; plants that have not been transformed will die. In some instances markers for selection are removed by backcrossing with the parent plant prior to commercial release.

The construct can be inserted in the plant genome by genetic recombination using the bacteria *Agrobacterium tumefaciens* or *A. rhizogenes*, or by direct methods like the gene gun or microinjection. Using plant viruses to insert genetic constructs into plants is also a possibility, but the technique is limited by the host range of the virus. For example, Cauliflower mosaic virus (CaMV) only infects cauliflower and related species. Another limitation of viral vectors is that the virus is not usually passed on the progeny, so every plant has to be inoculated.

The majority of commercially released transgenic plants are currently limited to plants that have introduced resistance to insect pests and herbicides. Insect resistance is achieved through incorporation of a gene from *Bacillus thuringiensis* (Bt) that encodes a protein that is toxic to some insects. For example, the cotton bollworm, a common cotton pest, feeds on Bt cotton it will ingest the toxin and die. Herbicides usually work by binding to certain plant enzymes and inhibiting their action. The enzymes that the herbicide inhibits are known as the herbicides *target site*. Herbicide resistance can be engineered into crops by expressing a version of *target site* protein that is not inhibited by the herbicide. This is the method used to produce glyphosate resistant crop plants.

Genetic modification of plants that can produce pharmaceuticals (and industrial chemicals), sometimes called *pharming*, is a rather radical new area of plant breeding.

Issues and Concerns

Modern plant breeding, whether classical or through genetic engineering, comes with issues of concern, particularly with regard to food crops. The question of whether breeding can have a negative effect on nutritional value is central in this respect. Although relatively little direct research in this area has been done, there are scientific indications that, by favoring certain aspects of a plant's development, other aspects may be retarded. A study published in the *Journal of the American College of Nutrition* in 2004, entitled *Changes in USDA Food Composition Data for 43 Garden Crops, 1950 to 1999*, compared nutritional analysis of vegetables done in 1950 and in 1999, and found substantial decreases in six of 13 nutrients measured, including 6% of protein and 38% of riboflavin. Reductions in calcium, phosphorus, iron and ascorbic acid were also found. The study, conducted at the Biochemical Institute, University of Texas at Austin, concluded in summary: *"We suggest that any real declines are generally most easily explained by changes in cultivated varieties between 1950 and 1999, in which there may be trade-offs between yield and nutrient content."*

The debate surrounding genetically modified food during the 1990s peaked in 1999 in terms of media coverage and risk perception, and continues today - for example, "*Germany has thrown its weight behind a growing European mutiny over genetically modified crops by banning the planting of a widely grown pest-resistant corn variety.*" The debate encompasses the ecological impact of genetically modified plants, the safety of genetically modified food and concepts used for safety evaluation like substantial equivalence. Such concerns are not new to plant breeding. Most countries have regulatory processes in place to help ensure that new crop varieties entering the marketplace are both safe and meet farmers' needs. Examples include variety registration, seed schemes, regulatory authorizations for GM plants, etc.

Plant breeders' rights is also a major and controversial issue. Today, production of new varieties is dominated by commercial plant breeders, who seek to protect their work and collect royalties through national and international agreements based in intellectual property rights. The range of related issues is complex. In the simplest terms, critics of the increasingly restrictive regulations argue that, through a combination of technical and economic pressures, commercial breeders are reducing biodiversity and significantly constraining individuals (such as farmers) from developing and trading seed on a regional level. Efforts to strengthen breeders' rights, for example, by lengthening periods of variety protection, are ongoing.

When new plant breeds or cultivars are bred, they must be maintained and propagated. Some plants are propagated by asexual means while others are propagated by seeds. Seed propagated cultivars require specific control over seed source and production procedures to maintain the integrity of the plant breeds results. Isolation is necessary to prevent cross contamination with related plants or the mixing of seeds after harvesting. Isolation is normally accomplished by planting distance but in certain crops, plants are enclosed in greenhouses or cages (most commonly used when producing F1 hybrids.)

Role of Plant Breeding in Organic Agriculture

Critics of organic agriculture claim it is too low-yielding to be a viable alternative to conventional agriculture. However, part of that poor performance may be the result of growing poorly adapted varieties. It is estimated that over 95% of organic agriculture is based on conventionally adapted varieties, even though the production environments found in organic vs. conventional farming systems are vastly different due to their distinctive management practices. Most notably, organic farmers have fewer inputs available than conventional growers to control their production environments. Breeding varieties specifically adapted to the unique conditions of organic agriculture is critical for this sector to realize its full potential. This requires selection for traits such as:

- Water use efficiency.
- Nutrient use efficiency (particularly nitrogen and phosphorus).
- Weed competitiveness.
- Tolerance of mechanical weed control.
- Pest/disease resistance.
- Early maturity (as a mechanism for avoidance of particular stresses).
- Abiotic stress tolerance (i.e. drought, salinity, etc).

Currently, few breeding programs are directed at organic agriculture and until recently those that did address this sector have generally relied on indirect selection (i.e. selection in conventional environments for traits considered important for organic agriculture). However, because the difference between organic and conventional environments is large, a given genotype may perform very differently in each environment due to an interaction between genes and the environment. If this interaction is severe enough, an important trait required for the organic environment may not be revealed in the conventional environment, which can result in the selection of poorly adapted individuals. To ensure the most adapted varieties are identified, advocates of organic breeding now promote the use of direct selection (i.e. selection in the target environment) for many agronomic traits.

There are many classical and modern breeding techniques that can be utilized for crop improvement in organic agriculture despite the ban on genetically modified organisms. For instance, controlled crosses between individuals allow desirable genetic variation to be recombined and transferred to seed progeny via natural processes. Marker assisted selection can also be employed as a diagnostics tool to facilitate selection of progeny who possess the desired trait(s), greatly speeding up the breeding process. This technique has proven particularly useful for the introgression of resistance genes into new backgrounds, as well as the efficient selection of many resistance genes pyramided into a single individual. Unfortunately, molecular markers are not currently available for many important traits, especially complex ones controlled by many genes.

Addressing Global Food Security Through Plant Breeding

For future agriculture to thrive there are necessary changes which must be made in accordance to arising global issues. These issues are arable land, harsh cropping conditions and food security which involves, being able to provide the world population with food containing sufficient nutrients. These crops need to be able to mature in several environments allowing for worldwide access, this is involves issues such as drought tolerance. These global issues are achievable through the process of plant breeding, as it offers the ability to select specific genes allowing the crop to perform at a level which yields the desired results.

Minimal Land Degradation

Land degradation is a major issue, as it can negatively impact the capability of the land to be productive. Poor agricultural management has a huge impact on the degradation of soil worldwide and it is Africa and Asia that are most affected. Through education and development of modified plants, these statistics can be reduced and agricultural land can become more productive. Plant breeding allows for an increase in yield with out the extra strain on the land. The genetically modified, Bt white maize, was introduced to South Africa and was surveyed in 33 large commercial farms and 368 small landholders properties and in both cases a higher yield was recorded.

Increased Yield Without Expansion

With an increasing population, the production of food needs to increase with it. It is estimated that a 70% increase in food production is needed by 2050 in order to meet the Declaration of the World Summit on Food Security. But with the natural degradation of agricultural land, simply planting more crops is no longer a viable option. Therefore, new varieties of plants need to be developed through plant breeding that generates an increase of yield without relying on an increase in land area. An example of this can be seen in Asia, where food production per capita has increased two-fold. This has been achieved through not only the use of fertilisers, but through the use of better crops that have been specifically designed for the area.

Breeding for Increased Nutritional Value

Plant breeding can contribute to global food security as it is a cost-effective tool for increasing nutritional value of forage and crops. Improvements in nutritional value for forage crops from the use of analytical chemistry and rumen fermentation technology have been recorded since 1960; this science and technology gave breeders the ability to screen thousands of samples within a small amount of time, meaning breeders could identify a high performing hybrid quicker. The main area genetic increases were made was in vitro dry matter digestibility (IVDMD) resulting in 0.7-2.5% increase, at just 1% increase in IVDMD a single Bos Taurus also known as beef cattle reported 3.2% increase in daily gains. This improvement indicates plant breeding is an essential tool in gearing future agriculture to perform at a more advanced level.

Breeding for Tolerance

Plant breeding of hybrid crops has become extremely popular worldwide in an effort to combat the harsh environment. With long periods of drought and lack of water or nitrogen stress tolerance has become a significant part of agriculture. Plant breeders have focused on identifying crops which will ensure crops perform under these conditions; a way to achieve this is finding strains of the crop that is resistance to drought conditions with low nitrogen. It is evident from this that plant breeding is vital for future agriculture to survive as it enables farmers to produce stress resistant crops hence improving food security.

Participatory Plant Breeding

The development of agricultural science, with phenomenon like the Green Revolution arising, have left millions of farmers in developing countries, most of whom operate small farms under unstable and difficult growing conditions, in a precarious situation. The adoption of new plant varieties by this group has been hampered by the constraints of poverty and the international policies promoting an industrialized model of agriculture. Their response has been the creation of a novel and promising set of research methods collectively known as participatory plant breeding. Participatory means that farmers are more involved in the breeding process and breeding goals are defined by farmers instead of international seed companies with their large-scale breeding programs. Farmers' groups and NGOs, for example, may wish to affirm local people's rights over genetic resources, produce seeds themselves, build farmers' technical expertise, or develop new products for niche markets, like organically grown food.

Plant Propagation

Plant propagation is the process of creating new plants from a variety of sources: seeds, cuttings, bulbs and other plant parts. Plant propagation can also refer to the artificial or natural dispersal of plants.

Gentian seedlings in a plant nursery.

Sexual Propagation (Seed)

One way to propagate an avocado seed.

Seeds and spores can be used for reproduction (through e.g. sowing). Seeds are typically produced from sexual reproduction within a species, because genetic recombination has occurred. A plant grown from seeds may have different characteristics from its parents. Some species produce seeds that require special conditions to germinate, such as cold treatment. The seeds of many Australian plants and plants from southern Africa and the American west require smoke or fire to germinate. Some plant species, including many trees do not produce seeds until they reach maturity, which may take many years. Seeds can be difficult to acquire and some plants do not produce seed at all. Some plants (like certain F1/F2 hybrids and GMO plants) may produce seed, but not fertile seed. In certain cases (like with GMO's), this is done to prevent the accidental spreading of these plants (which are generally non-native crops), for example by birds and other animals.

Asexual Propagation

Softwood stemcuttings rooting in a controlled environment.

Plants have a number of mechanisms for asexual or vegetative reproduction. Some of these have been taken advantage of by horticulturists and gardeners to multiply or clone plants rapidly. People also use methods that plants do not use, such as tissue culture and grafting. Plants are produced using material from a single parent and as such there is no exchange of genetic material, therefore vegetative propagation methods almost always produce plants that are identical to the parent. Vegetative reproduction uses plants parts such as roots, stems and leaves. In some plants seeds can be produced without fertilization and the seeds contain only the genetic material of the parent plant. Therefore, propagation via asexual seeds or apomixis is asexual reproduction but not vegetative propagation.

Techniques for vegetative propagation include:

- Air or ground layering,
- Division,
- Grafting and bud grafting, widely used in fruit tree propagation,
- Micropropagation,
- Stolons or runners,
- Storage organs such as bulbs, corms, tubers and rhizomes,
- Striking or cuttings,
- Twin-scaling,
- Offsets.

Heated Propagator

A heated propagator is a horticultural device to maintain a warm and damp environment for seeds and cuttings to grow in.

This can be in the form of a clear enclosed bin sitting over a hotpad, or even a portable heater pointed at the bin. The key is to keep the moisture in the clear bin, while keeping lighting over the top of it, usually.

Seed Propagation Mat

An electric seed-propagation mat is a heated rubber mat covered by a metal cage which is used in gardening. The mats are made so that planters containing seedlings can be placed on top of the metal cage without the risk of starting a fire. In extreme cold, gardeners place a loose plastic cover over the planters/mats which creates a sort of miniature greenhouse. The constant and predictable heat allows people to garden in the winter months when the weather is generally too cold for seedlings to survive naturally. When combined with a lighting system, many plants can be grown indoors using these mats.

Plant Morphology

Plant morphology or phytomorphology is the study of the physical form and external structure of plants. This is usually considered distinct from plant anatomy, which is the study of the internal structure of plants, especially at the microscopic level. Plant morphology is useful in the visual identification of plants.

Inflorescences emerging from protective coverings.

Scope

Plant morphology "represents a study of the development, form, and structure of plants, and, by implication, an attempt to interpret these on the basis of similarity of plan and origin." There are four major areas of investigation in plant morphology, and each overlaps with another field of the biological sciences.

Looking up into the branch structure of a *Pinus sylvestris* tree.

Asclepias syriaca showing complex morphology of the flowers.

First of all, morphology is comparative, meaning that the morphologist examines structures in many different plants of the same or different species, then draws comparisons and formulates ideas about similarities. When structures in different species are believed to exist and develop as a result of common, inherited genetic pathways, those structures are termed homologous. For example, the leaves of pine, oak, and cabbage all look very different, but share certain basic structures and arrangement of parts. The homology of leaves is an easy conclusion to make. The plant morphologist goes further, and discovers that the spines of cactus also share the same basic structure and development as leaves in other plants, and therefore cactus spines are homologous to leaves as well. This aspect of plant morphology overlaps with the study of plant evolution and paleobotany.

Secondly, plant morphology observes both the vegetative (somatic) structures of plants, as well as the reproductive structures. The vegetative structures of vascular plants includes the study of the shoot system, composed of stems and leaves, as well as the root system. The reproductive structures are more varied, and are usually specific to a particular group of plants, such as flowers and seeds, fern sori, and moss capsules. The detailed study of reproductive structures in plants led to the discovery of the alternation of generations found in all plants and most algae. This area of plant morphology overlaps with the study of biodiversity and plant systematics.

Thirdly, plant morphology studies plant structure at a range of scales. At the smallest scales are ultrastructure, the general structural features of cells visible only with the aid of an electron microscope, and cytology, the study of cells using optical microscopy. At this scale, plant morphology overlaps with plant anatomy as a field of study. At the largest scale is the study of plant growth habit, the overall architecture of a plant. The pattern of branching in a tree will vary from species to species, as will the appearance of a plant as a tree, herb, or grass.

Fourthly, plant morphology examines the pattern of development, the process by which structures originate and mature as a plant grows. While animals produce all the body parts they will ever have from early in their life, plants constantly produce new tissues and structures throughout their life. A living plant always has embryonic tissues. The way in which new structures mature as they are produced may be affected by the point in the plant's life when they begin to develop, as well as by the environment to which the structures are exposed. A morphologist studies this process, the causes, and its result. This area of plant morphology overlaps with plant physiology and ecology.

A Comparative Science

A plant morphologist makes comparisons between structures in many different plants of the same or different species. Making such comparisons between similar structures in different plants tackles the question of *why* the structures are similar. It is quite likely that similar underlying causes of genetics, physiology, or response to the environment have led to this similarity in appearance. The result of scientific investigation into these causes can lead to one of two insights into the underlying biology:

1. Homology - the structure is similar between the two species because of shared ancestry and common genetics.

2. Convergence - the structure is similar between the two species because of independent adaptation to common environmental pressures.

Understanding which characteristics and structures belong to each type is an important part of understanding plant evolution. The evolutionary biologist relies on the plant morphologist to interpret structures, and in turn provides phylogenies of plant relationships that may lead to new morphological insights.

Homology

When structures in different species are believed to exist and develop as a result of common, inherited genetic pathways, those structures are termed *homologous*. For example, the leaves of pine, oak, and cabbage all look very different, but share certain basic structures and arrangement of parts. The homology of leaves is an easy conclusion to make. The plant morphologist goes further, and discovers that the spines of cactus also share the same basic structure and development as leaves in other plants, and therefore cactus spines are homologous to leaves as well.

Convergence

When structures in different species are believed to exist and develop as a result of common adaptive responses to environmental pressure, those structures are termed *convergent*. For example, the fronds of *Bryopsis plumosa* and stems of *Asparagus setaceus* both have the same feathery branching appearance, even though one is an alga and one is a flowering plant. The similarity in overall structure occurs independently as a result of convergence. The growth form of many cacti and species of *Euphorbia* is very similar, even though they belong to widely distant families. The similarity results from common solutions to the problem of surviving in a hot, dry environment.

Astrophytum asterias, a cactus.

Euphorbia obesa, a spurge.

Vegetative and Reproductive Characteristics

Plant morphology treats both the vegetative structures of plants, as well as the reproductive structures.

The vegetative (somatic) structures of vascular plants include two major organ systems: (1) a shoot system, composed of stems and leaves, and (2) a root system. These two systems are common to nearly all vascular plants, and provide a unifying theme for the study of plant morphology.

By contrast, the reproductive structures are varied, and are usually specific to a particular group of plants. Structures such as flowers and fruits are only found in the angiosperms; sori are only found in ferns; and seed cones are only found in conifers and other gymnosperms. Reproductive characters are therefore regarded as more useful for the classification of plants than vegetative characters.

Use in Identification

Plant biologists use morphological characters of plants which can be compared, measured, counted and described to assess the differences or similarities in plant taxa and use these characters for plant identification, classification and descriptions.

When characters are used in descriptions or for identification they are called diagnostic or key characters which can be either qualitative and quantitative.

1. Quantitative characters are morphological features that can be counted or measured for example a plant species has flower petals 10–12 mm wide.
2. Qualitative characters are morphological features such as leaf shape, flower color or pubescence.

Both kinds of characters can be very useful for the identification of plants.

Alternation of Generations

The detailed study of reproductive structures in plants led to the discovery of the alternation of generations, found in all plants and most algae, by the German botanist Wilhelm Hofmeister. This discovery is one of the most important made in all of plant morphology, since it provides a common basis for understanding the life cycle of all plants.

Pigmentation in Plants

The primary function of pigments in plants is photosynthesis, which uses the green pigment chlorophyll along with several red and yellow pigments that help to capture as much light energy as possible. Pigments are also an important factor in attracting insects to flowers to encourage pollination.

Plant pigments include a variety of different kinds of molecule, including porphyrins, carotenoids, anthocyanins and betalains. All biological pigments selectively absorb certain wavelengths of light while reflecting others. The light that is absorbed may be used by the plant to power chemical reactions, while the reflected wavelengths of light determine the color the pigment will appear to the eye.

Morphology in Development

Plant development is the process by which structures originate and mature as a plant grows. It is a subject studies in plant anatomy and plant physiology as well as plant morphology.

The process of development in plants is fundamentally different from that seen in vertebrate animals. When an animal embryo begins to develop, it will very early produce all of the body parts that it will ever have in its life. When the animal is born (or hatches from its egg), it has all its body parts and from that point will only grow larger and more mature. By contrast, plants constantly produce new tissues and structures throughout their life from meristems located at the tips of organs, or between mature tissues. Thus, a living plant always has embryonic tissues.

The properties of organization seen in a plant are emergent properties which are more than the sum of the individual parts. "The assembly of these tissues and functions into an integrated multicellular organism yields not only the characteristics of the separate parts and processes but also quite a new set of characteristics which would not have been predictable on the basis of examination of the separate parts." In other words, knowing everything about the molecules in a plant are not enough to predict characteristics of the cells; and knowing all the properties of the cells will not predict all the properties of a plant's structure.

Growth

A vascular plant begins from a single celled zygote, formed by fertilisation of an egg cell by a sperm cell. From that point, it begins to divide to form a plant embryo through the process of embryogenesis. As this happens, the resulting cells will organize so that one end becomes the first root, while the other end forms the tip of the shoot. In seed plants, the embryo will develop one or more "seed leaves" (cotyledons). By the end of embryogenesis, the young plant will have all the parts necessary to begin in its life.

Once the embryo germinates from its seed or parent plant, it begins to produce additional organs (leaves, stems, and roots) through the process of organogenesis. New roots grow from root meristems located at the tip of the root, and new stems and leaves grow from shoot meristems located at the tip of the shoot. Branching occurs when small clumps of cells left behind by the meristem, and which have not yet undergone cellular differentiation to form a specialized tissue, begin to grow as the tip of a new root or shoot. Growth from any such meristem at the tip of a root or shoot is termed primary growth and results in the lengthening of that root or shoot. Secondary growth results in widening of a root or shoot from divisions of cells in a cambium.

In addition to growth by cell division, a plant may grow through cell elongation. This occurs when individual cells or groups of cells grow longer. Not all plant cells will grow to the same length. When cells on one side of a stem grow longer and faster than cells on the other side, the stem will bend to the side of the slower growing cells as a result. This directional growth can occur via a plant's response to a particular stimulus, such as light (phototropism), gravity (gravitropism), water, (hydrotropism), and physical contact (thigmotropism).

Plant growth and development are mediated by specific plant hormones and plant growth regulators (PGRs) (Ross et al. 1983). Endogenous hormone levels are influenced by plant age, cold hardiness, dormancy, and other metabolic conditions; photoperiod, drought, temperature, and other external environmental conditions; and exogenous sources of PGRs, e.g., externally applied and of rhizospheric origin.

Morphological Variation

Plants exhibit natural variation in their form and structure. While all organisms vary from individual to individual, plants exhibit an additional type of variation. Within a single individual, parts are repeated which may differ in form and structure from other similar parts. This variation is most easily seen in the leaves of a plant, though other organs such as stems and flowers may show similar variation. There are three primary causes of this variation: positional effects, environmental effects, and juvenility.

Evolution of Plant Morphology

Transcription factors and transcriptional regulatory networks play key roles in plant morphogenesis and their evolution. During plant landing, many novel transcription factor families emerged and are preferentially wired into the networks of multicellular development, reproduction, and organ development, contributing to more complex morphogenesis of land plants.

Positional Effects

Variation in leaves from the giant ragweed illustrating positional effects. The lobed leaves come from the base of the plant, while the unlobed leaves come from the top of the plant.

Although plants produce numerous copies of the same organ during their lives, not all copies of a particular organ will be identical. There is variation among the parts of a mature plant resulting from the relative position where the organ is produced. For example, along a new branch the leaves may vary in a consistent pattern along the branch. The form of leaves produced near the base of the branch will differ from leaves produced at the tip of the plant, and this difference is consistent from branch to branch on a given plant and in a given species. This difference persists after the leaves at both ends of the branch have matured, and is not the result of some leaves being younger than others.

Environmental Effects

The way in which new structures mature as they are produced may be affected by the point in the plants life when they begin to develop, as well as by the environment to which the structures are exposed. This can be seen in aquatic plants and emergent plants.

Temperature

Temperature has a multiplicity of effects on plants depending on a variety of factors, including the size and condition of the plant and the temperature and duration of exposure. The smaller and more succulent the plant, the greater the susceptibility to damage or death from temperatures that are too high or too low. Temperature affects the rate of biochemical and physiological processes, rates generally (within limits) increasing with temperature. However, the Van't Hoff relationship for monomolecular reactions (which states that the velocity of a reaction is doubled or trebled by a temperature increase of 10 °C) does not strictly hold for biological processes, especially at low and high temperatures.

When water freezes in plants, the consequences for the plant depend very much on whether the freezing occurs intracellularly (within cells) or outside cells in intercellular (extracellular) spaces. Intracellular freezing usually kills the cell regardless of the hardiness of the plant and its tissues. Intracellular freezing seldom occurs in nature, but moderate rates of decrease in temperature, e.g., 1 °C to 6 °C/hour, cause intercellular ice to form, and this "extraorgan ice" may or may not be lethal, depending on the hardiness of the tissue.

At freezing temperatures, water in the intercellular spaces of plant tissues freezes first, though the water may remain unfrozen until temperatures fall below 7 °C. After the initial formation of ice intercellularly, the cells shrink as water is lost to the segregated ice. The cells undergo freeze-drying, the dehydration being the basic cause of freezing injury.

The rate of cooling has been shown to influence the frost resistance of tissues, but the actual rate of freezing will depend not only on the cooling rate, but also on the degree of supercooling and the properties of the tissue. Sakai (1979a) demonstrated ice segregation in shoot primordia of Alaskan white and black spruces when cooled slowly to 30 °C to -40 °C. These freeze-dehydrated buds survived immersion in liquid nitrogen when slowly rewarmed. Floral primordia responded similarly. Extraorgan freezing in the primordia accounts for the ability of the hardiest of the boreal conifers to survive winters in regions when air temperatures often fall to -50 °C or lower. The hardiness of the winter buds of such conifers is enhanced by the smallness of the buds, by the evolution of faster translocation of water, and an ability to tolerate intensive freeze dehydration. In boreal species of

Picea and *Pinus*, the frost resistance of 1-year-old seedlings is on a par with mature plants, given similar states of dormancy.

Juvenility

The organs and tissues produced by a young plant, such as a seedling, are often different from those that are produced by the same plant when it is older. This phenomenon is known as juvenility or heteroblasty. For example, young trees will produce longer, leaner branches that grow upwards more than the branches they will produce as a fully grown tree. In addition, leaves produced during early growth tend to be larger, thinner, and more irregular than leaves on the adult plant. Specimens of juvenile plants may look so completely different from adult plants of the same species that egg-laying insects do not recognize the plant as food for their young. Differences are seen in rootability and flowering and can be seen in the same mature tree. Juvenile cuttings taken from the base of a tree will form roots much more readily than cuttings originating from the mid to upper crown. Flowering close to the base of a tree is absent or less profuse than flowering in the higher branches especially when a young tree first reaches flowering age.

Juvenility in a seedling of European beech. There is a marked difference in shape between the first dark green "seed leaves" and the lighter second pair of leaves.

The transition from early to late growth forms is referred to as 'vegetative phase change', but there is some disagreement about terminology.

Some Recent Developments

Rolf Sattler has revised fundamental concepts of comparative morphology such as the concept of homology. He emphasized that homology should also include partial homology and quantitative homology. This leads to a continuum morphology that demonstrates a continuum between the morphological categories of root, shoot, stem (caulome), leaf (phyllome), and hair (trichome). How intermediates between the categories are best described has been discussed by Bruce K. Kirchoff et al.

Honoring Agnes Arber, author of the partial-shoot theory of the leaf, Rutishauser and Isler called the continuum approach Fuzzy Arberian Morphology (FAM). "Fuzzy" refers to fuzzy logic, "Arberian" to Agnes Arber. Rutishauser and Isler emphasized that this approach is not only supported by many morphological data but also by evidence from molecular genetics. More recent evidence from molecular genetics provides further support for continuum morphology. James (2009) concluded that "it is now widely accepted that... radiality [characteristic of most shoots] and dorsiventrality [characteristic of leaves] are but extremes of a continuous spectrum. In fact, it is simply the timing of the KNOX gene expression!." Eckardt and Baum (2010) concluded that "it is now generally accepted that compound leaves express both leaf and shoot properties".

Process morphology (dynamic morphology) describes and analyzes the dynamic continuum of plant form. According to this approach, structures do not *have* process(es), they *are* process(es). Thus, the structure/process dichotomy is overcome by "an enlargement of our concept of 'structure' so as to include and recognize that in the living organism it is not merely a question of spatial structure with an 'activity' as something over or against it, but that the concrete organism is a spatio-*temporal* structure and that this spatio-temporal structure is the activity itself."

For Jeune, Barabé and Lacroix, classical morphology (that is, mainstream morphology, based on a qualitative homology concept implying mutually exclusive categories) and continuum morphology are sub-classes of the more encompassing process morphology (dynamic morphology).

Plant Ecology

Plant ecology is a subdiscipline of ecology which studies the distribution and abundance of plants, the effects of environmental factors upon the abundance of plants, and the interactions among and between plants and other organisms. Examples of these are the distribution of temperate deciduous forests in North America, the effects of drought or flooding upon plant survival, and competition among desert plants for water, or effects of herds of grazing animals upon the composition of grasslands.

A tropical plant community on Diego Garcia.

A global overview of the Earth's major vegetation types is provided by O.W. Archibold. He recognizes 11 major vegetation types: tropical forests, tropical savannas, arid regions (deserts), Mediterranean ecosystems, temperate forest ecosystems, temperate grasslands, coniferous forests, tundra (both polar and high mountain), terrestrial wetlands, freshwater ecosystems and coastal/marine systems. This breadth of topics shows the complexity of plant ecology, since it includes plants from floating single-celled algae up to large canopy forming trees.

One feature that defines plants is photosynthesis. One of the most important aspects of plant ecology is the role plants have played in creating the oxygenated atmosphere of earth, an event that occurred some 2 billion years ago. It can be dated by the deposition of banded iron formations, distinctive sedimentary rocks with large amounts of iron oxide. At the same time, plants began removing carbon dioxide from the atmosphere, thereby initiating the process of controlling Earth's climate. A long term trend of the Earth has been toward increasing oxygen and decreasing carbon dioxide, and many other events in the Earths history, like the first movement of life onto land, are likely tied to this sequence of events.

One of the early classic books on plant ecology was written by J.E. Weaver and F.E. Clements. It talks broadly about plant communities, and particularly the importance of forces like competition and processes like succession. Although some of the terminology is dated, this important book can still often be obtained in used book stores.

Plant ecology can also be divided by levels of organization including plant ecophysiology, plant population ecology, community ecology, ecosystem ecology, landscape ecology and biosphere ecology.

The study of plants and vegetation is complicated by their form. First, most plants are rooted in the soil, which makes it difficult to observe and measure nutrient uptake and species interactions. Second, plants often reproduce vegetatively, that is asexually, in a way that makes it difficult to distinguish individual plants. Indeed, the very concept of an individual is doubtful, since even a tree may be regarded as a large collection of linked meristems. Hence, plant ecology and animal ecology have different styles of approach to problems that involve processes like reproduction, dispersal and mutualism. Some plant ecologists have placed considerable emphasis upon trying to treat plant populations as if they were animal populations, focusing on population ecology. Many other ecologists believe that while it is useful to draw upon population ecology to solve certain scientific problems, plants demand that ecologists work with multiple perspectives, appropriate to the problem, the scale and the situation.

History

Alexander von Humboldt's work connecting plant distributions with environmental factors played an important role in the genesis of the discipline of plant ecology.

Plant ecology has its origin in the application of plant physiology to the questions raised by plant geographers. Carl Ludwig Willdenow was one of the first to note that similar climates produced similar types of vegetation, even when they were located in different parts of the world. Willdenow's student, Alexander von Humboldt, used physiognomy to describe vegetation types and observed that the distribution vegetation types was based on environmental factors. Later plant geographers who built upon Humboldt's work included Joakim Frederik Schouw, A.P. de Candolle,

August Grisebach and Anton Kerner von Marilaun. Schouw's work, published in 1822, linked plant distributions to environmental factors (especially temperature) and established the practice of naming plant associations by adding the suffix *-etum* to the name of the dominant species. Working from herbarium collections, De Candolle searched for general rules of plant distribution and settled on using temperature as well. Grisebach's two-volume work, *Die Vegetation der Erde nach Ihrer Klimatischen Anordnung*, published in 1872, saw plant geography reach its "ultimate form" as a descriptive field.

Starting in the 1870s, Swiss botanist Simon Schwendener, together with his students and colleagues, established the link between plant morphology and physiological adaptations, laying the groundwork for the first ecology textbooks, Eugenius Warming's *Plantesamfund* (published in 1895) and Andreas Schimper's 1898 *Pflanzengeographie auf Physiologischer Grundlage*. Warming successfully incorporated plant morphology, physiology taxonomy and biogeography into plant geography to create the field of plant ecology. Although more morphological than physiological, Schimper's has been considered the beginning of plant physiological ecology. Plant ecology was initially built around static ideas of plant distribution; incorporating the concept of succession added an element to change through time to the field. Henry Chandler Cowles' studies of plant succession on the Lake Michigan sand dunes (published in 1899) and Frederic Clements' 1916 monograph on the subject established it as a key element of plant ecology.

Plant ecology developed within the wider discipline of ecology over the twentieth century. Inspired by Warming's *Plantesamfund*, Arthur Tansley set out to map British plant communities. In 1904 he teamed up with William Gardner Smith and others involved in vegetation mapping to establish the Central Committee for the Survey and Study of British Vegetation, later shortened to British Vegetation Committee. In 1913, the British Vegetation Committee organised the British Ecological Society (BES), the first professional society of ecologists. This was followed in 1917 by the establishment of the Ecological Society of America (ESA); plant ecologists formed the largest subgroup among the inaugural members of the ESA.

Cowles' students played an important role in the development of the field of plant ecology during the first half of the twentieth century, among them William S. Cooper, E. Lucy Braun and Edgar Transeau.

Graphical timeline of plant ecologists

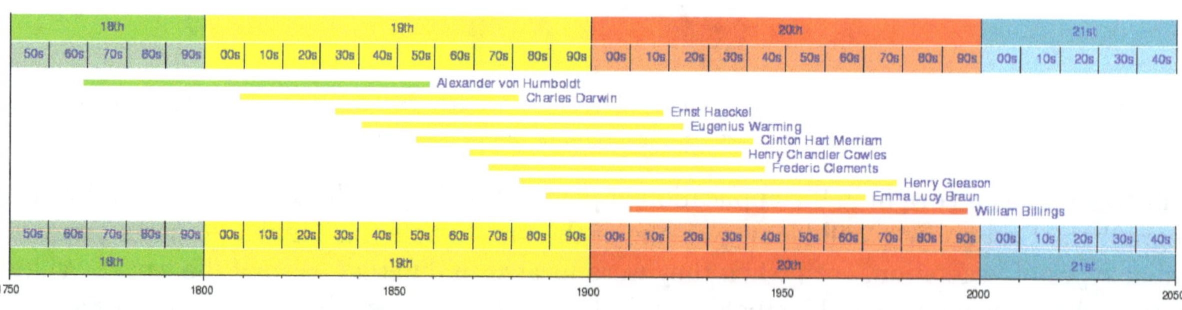

Distribution

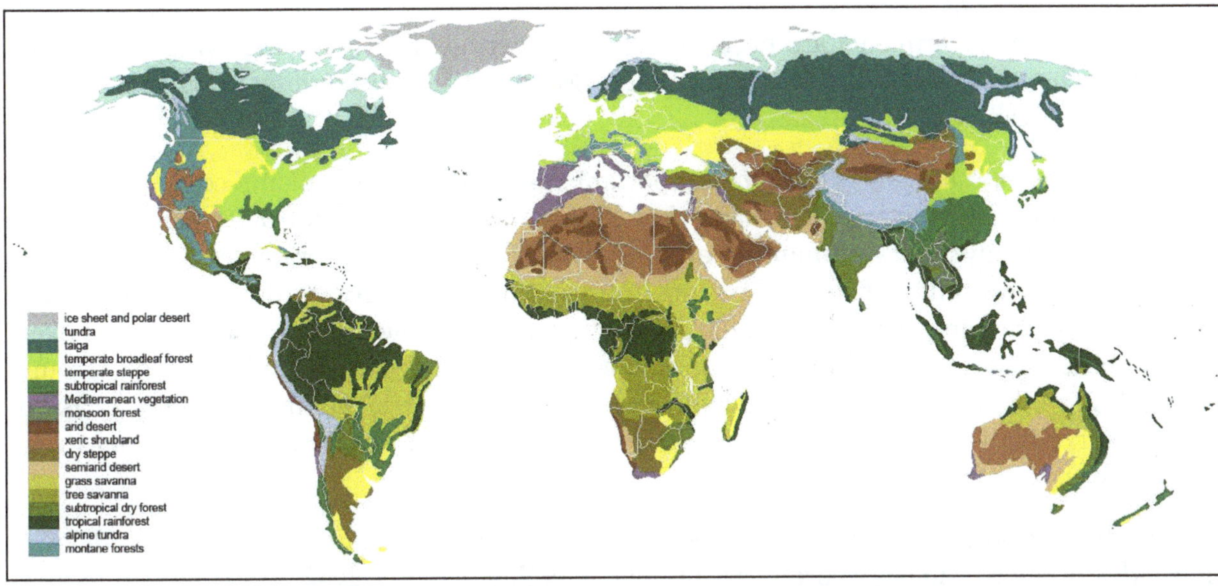

World biomes are based upon the type of dominant plant.

Plant distributions is governed by a combination of historical factors, ecophysiology and biotic interactions. The set of species that can be present at a given site is limited by historical contingency. In order to show up, a species must either have evolved in an area or dispersed there (either naturally or through human agency), and must not have gone locally extinct. The set of species present locally is further limited to those that possess the physiological adaptations to survive the environmental conditions that exist. This group is further shaped through interactions with other species.

Plant communities are broadly distributed into biomes based on the form of the dominant plant species. For example, grasslands are dominated by grasses, while forests are dominated by trees. Biomes are determined by regional climates, mostly temperature and precipitation, and follow general latitudinal trends. Within biomes, there may be many ecological communities, which are impacted not only by climate and a variety of smaller-scale features, including soils, hydrology, and disturbance regime. Biomes also change with elevation, high elevations often resembling those found at higher latitudes.

Biological Interactions

Competition

Plants, like most life forms, require relatively few basic elements: carbon, hydrogen, oxygen, nitrogen, phosphorus and sulphur; hence they are known as CHNOPS life forms. There are also lesser elements needed as well, frequently termed micronutrients, such as magnesium and sodium. When plants grow in close proximity, they may deplete supplies of these elements and have a negative impact upon neighbours. Competition for resources vary from complete symmetric (all individuals receive the same amount of resources, irrespective of their size) to perfectly size symmetric (all individuals exploit the same amount of resource per unit biomass) to absolutely size-asymmetric (the largest individuals exploit all the available resource). The degree of size asymmetry has major effects on the structure and diversity of ecological communities.In many cases (perhaps most) the negative effects upon neighbours arise from size asymmetric competition for light. In other cases, there may be competition below ground for water, nitrogen, or phosphorus. To detect and measure competition, experiments are necessary; these experiments require removing neighbours, and measuring responses in the remaining plants. Many such studies are required before useful generalizations can be drawn.

Overall, it appears that light is the most important resource for which plants compete, and the increase in plant height over evolutionary time likely reflects selection for taller plants to better intercept light. Many plant communities are therefore organized into hierarchies based upon the relative competitive abilities for light. In some systems, particularly infertile or arid systems, below ground competition may be more significant. Along natural gradients of soil fertility, it is likely that the ratio of above ground to below ground competition changes, with higher above ground competition in the more fertile soils. Plants that are relatively weak competitors may escape in time (by surviving as buried seeds) or in space (by dispsersing to a new location away from strong competitors).

In principle, it is possible to examine competition at the level of the limiting resources if a detailed knowledge of the physiological processes of the competing plants is available. However, in most terrestrial ecological studies, there is only little information on the uptake and dynamics of the resources that limit the growth of different plant species, and, instead, competition is inferred from observed negative effects of neighbouring plants without knowing precisely which resources the plants were competing for. In certain situations, plants may compete for a single growth-limiting resource, perhaps for light in agricultural systems with sufficient water and nutrients, or in dense stands of marsh vegetation, but in many natural ecosystems plants may be colimited by several resources, e.g. light, phosphorus and nitrogen at the same time.

Therefore, there are many details that remain to be uncovered, particularly the kinds of competition that arise in natural plant communities, the specific resource(s), the relative importance of different resources, and the role of other factors like stress or disturbance in regulating the importance of competition.

Mutualism

Mutualism is defined as an interaction "between two species or individuals that is beneficial to both". Probably the most widespread example in plants is the mutual beneficial relationship between plants and fungi, known as mycorrhizae. The plant is assisted with nutrient uptake, while the fungus receives carbohydrates. Some the earliest known fossil plants even have fossil mycorrhizae on their rhizomes.

The flowering plants are a group that have evolved by using two major mutualisms. First, flowers are pollinated by insects. This relationship seems to have its origins in beetles feeding on primitive flowers, eating pollen and also acting (unwittingly) as pollinators. Second, fruits are eaten by animals, and the animals then disperse the seeds. Thus, the flowering plants actually have three major types of mutualism, since most higher plants also have mycorrhizae.

Plants may also have beneficial effects upon one another, but this is less common. Examples might include "nurse plants" whose shade allows young cacti to establish. Most examples of mutualism, however, are largely beneficial to only one of the partners, and may not really be true mutualism. The term used for these more one-sided relationships, which are mostly beneficial to one participant, is facilitation. Facilitation among neighboring plants may act by reducing the negative impacts of a stressful environment. In general, facilitation is more likely to occur in physically stressful environments than in favorable environments, where competition may be the most important interaction among species.

Commensalism is similar to facilitation, in that one plant is mostly exploiting another. A familiar example is the ephiphytes which grow on branches of tropical trees, or even mosses which grow on trees in deciduous forests.

It is important to keep track of the benefits received by each species to determine the appropriate term. Although people are often fascinated by unusual examples, it is important to remember that in plants, the main mutualisms are mycorrhizae, pollination, and seed dispersal.

Herbivory

An important ecological function of plants is that they produce organic compounds for herbivores in the bottom of the food web. A large number of plant traits, from thorns to chemical defenses, can be related to the intensity of herbivory. Large herbivores can also have many effects on vegetation. These include removing selected species, creating gaps for regeneration of new individuals, recycling nutrients, and dispersing seeds. Certain ecosystem types, such as grasslands, may be dominated by the effects of large herbivores, although fire is also an equally important factor in this biome. In few cases, herbivores are capable of nearly removing all the vegetation at a site (for example, geese in the Hudson Bay Lowlands of Canada, and nutria in the marshes of Louisiana) but normally heribovres have a more selective impact, particularly when large predators control the abundance of herbivores. The usual method of studying the effects of herbivores is to build exclosures, where they cannot feed, and compare the plant communities in the exclosures to those outside over many years. Often such long term experiments show that herbivores have a significant effect upon the species that make up the plant community.

Other Topics

Abundance

The ecological success of a plant species in a specific environment may be quantified by its abundance, and depending on the life form of the plant different measures of abundance may be relevant, e.g. density, biomass, or plant cover.

The change in the abundance of a plant species may be due to both abiotic factors, e.g. climate change, or biotic factors, e.g. herbivory or interspecific competition.

Colonisation and Local Extinction

Whether a plant species is present at a local area depends on the processes of colonisation and local extinction. The probability of colonisation decreases with distance to neighboring habitats where the species is present and increases with plant abundance and fecundity in neighboring habitats and the dispersal distance of the species. The probability of local extinction decreases with abundance (both living plants and seeds in the soil seed bank).

Plant Physiology

Plant physiology is a subdiscipline of botany concerned with the functioning, or physiology, of plants. Closely related fields include plant morphology (structure of plants), plant ecology (interactions with the environment), phytochemistry (biochemistry of plants), cell biology, genetics, biophysics and molecular biology.

A germination rate experiment.

Fundamental processes such as photosynthesis, respiration, plant nutrition, plant hormone functions, tropisms, nastic movements, photoperiodism, photomorphogenesis, circadian rhythms, environmental stress physiology, seed germination, dormancy and stomata function and transpiration, both parts of plant water relations, are studied by plant physiologists.

Aims

The field of plant physiology includes the study of all the internal activities of plants—those chemical and physical processes associated with life as they occur in plants. This includes study at many levels of scale of size and time. At the smallest scale are molecular interactions of photosynthesis and internal diffusion of water, minerals, and nutrients. At the largest scale are the processes of plant development, seasonality, dormancy, and reproductive control. Major subdisciplines of plant physiology include phytochemistry (the study of the biochemistry of plants) and phytopathology (the study of disease in plants). The scope of plant physiology as a discipline may be divided into several major areas of research.

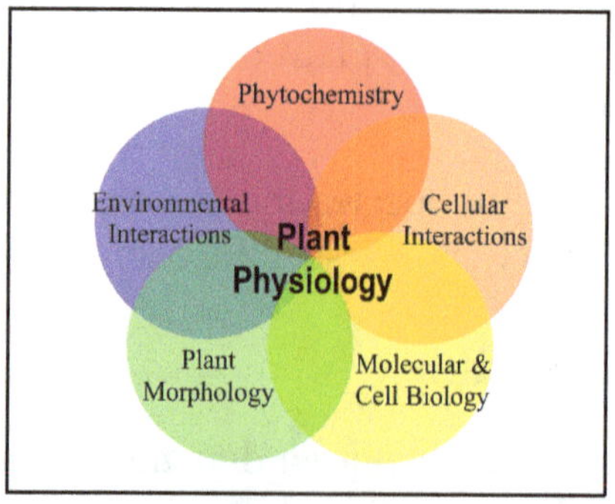

Five key areas of study within plant physiology.

First, the study of phytochemistry (plant chemistry) is included within the domain of plant physiology. To function and survive, plants produce a wide array of chemical compounds not found in other organisms. Photosynthesis requires a large array of pigments, enzymes, and other compounds to function. Because they cannot move, plants must also defend themselves chemically from herbivores, pathogens and competition from other plants. They do this by producing toxins and foul-tasting or smelling chemicals. Other compounds defend plants against disease, permit survival during drought, and prepare plants for dormancy, while other compounds are used to attract pollinators or herbivores to spread ripe seeds.

Secondly, plant physiology includes the study of biological and chemical processes of individual plant cells. Plant cells have a number of features that distinguish them from cells of animals, and which lead to major differences in the way that plant life behaves and responds differently from animal life. For example, plant cells have a cell wall which restricts the shape of plant cells and thereby limits the flexibility and mobility of plants. Plant cells also contain chlorophyll, a chemical compound that interacts with light in a way that enables plants to manufacture their own nutrients rather than consuming other living things as animals do.

Thirdly, plant physiology deals with interactions between cells, tissues, and organs within a plant. Different cells and tissues are physically and chemically specialized to perform different functions. Roots and rhizoids function to anchor the plant and acquire minerals in the soil. Leaves catch light in order to manufacture nutrients. For both of these organs to remain living, minerals that the roots acquire must be transported to the leaves, and the nutrients manufactured in the leaves must be transported to the roots. Plants have developed a number of ways to achieve this transport, such as vascular tissue, and the functioning of the various modes of transport is studied by plant physiologists.

Fourthly, plant physiologists study the ways that plants control or regulate internal functions. Like animals, plants produce chemicals called hormones which are produced in one part of the plant to signal cells in another part of the plant to respond. Many flowering plants bloom at the appropriate time because of light-sensitive compounds that respond to the length of the night, a phenomenon known as photoperiodism. The ripening of fruit and loss of leaves in the winter are controlled in part by the production of the gas ethylene by the plant.

Finally, plant physiology includes the study of plant response to environmental conditions and their variation, a field known as environmental physiology. Stress from water loss, changes in air chemistry, or crowding by other plants can lead to changes in the way a plant functions. These changes may be affected by genetic, chemical, and physical factors.

Biochemistry of Plants

The chemical elements of which plants are constructed—principally carbon, oxygen, hydrogen, nitrogen, phosphorus, sulfur, etc.—are the same as for all other life forms animals, fungi, bacteria and even viruses. Only the details of the molecules into which they are assembled differs.

Despite this underlying similarity, plants produce a vast array of chemical compounds with unique properties which they use to cope with their environment. Pigments are used by plants to absorb or detect light, and are extracted by humans for use in dyes. Other plant products may be used for the manufacture of commercially important rubber or biofuel. Perhaps the most celebrated compounds from plants are those with pharmacological activity, such as salicylic acid from which aspirin is made, morphine, and digoxin. Drug companies spend billions of dollars each year researching plant compounds for potential medicinal benefits.

Latex being collected from a tapped rubber tree.

Constituent Elements

Plants require some nutrients, such as carbon and nitrogen, in large quantities to survive. Such nutrients are termed macronutrients, where the prefix *macro-* (large) refers to the quantity needed, not the size of the nutrient particles themselves. Other nutrients, called micronutrients, are required only in trace amounts for plants to remain healthy. Such micronutrients are usually absorbed as ions dissolved in water taken from the soil, though carnivorous plants acquire some of their micronutrients from captured prey.

The following tables list element nutrients essential to plants. Uses within plants are generalized.

Macronutrients - necessary in large quantities		
Element	Form of uptake	Notes
Nitrogen	NO_3^-, NH_4^+	Nucleic acids, proteins, hormones, etc.
Oxygen	O_2 H_2O	Cellulose, starch, other organic compounds
Carbon	CO_2	Cellulose, starch, other organic compounds
Hydrogen	H_2O	Cellulose, starch, other organic compounds
Potassium	K^+	Cofactor in protein synthesis, water balance, etc.
Calcium	Ca^{2+}	Membrane synthesis and stabilization
Magnesium	Mg^{2+}	Element essential for chlorophyll
Phosphorus	$H_2PO_4^-$	Nucleic acids, phospholipids, ATP
Sulfur	SO_4^{2-}	Constituent of proteins

Micronutrients - necessary in small quantities		
Element	Form of uptake	Notes
Chlorine	Cl^-	Photosystem II and stomata function
Iron	Fe^{2+}, Fe^{3+}	Chlorophyll formation
Boron	HBO_3	Crosslinking pectin
Manganese	Mn^{2+}	Activity of some enzymes
Zinc	Zn^{2+}	Involved in the synthesis of enzymes and chlorophyll
Copper	Cu^+	Enzymes for lignin synthesis
Molybdenum	MoO_4^{2-}	Nitrogen fixation, reduction of nitrates
Nickel	Ni^{2+}	Enzymatic cofactor in the metabolism of nitrogen compounds

Pigments

Among the most important molecules for plant function are the pigments. Plant pigments include a variety of different kinds of molecules, including porphyrins, carotenoids, and anthocyanins. All biological pigments selectively absorb certain wavelengths of light while reflecting others. The light that is absorbed may be used by the plant to power chemical reactions, while the reflected wavelengths of light determine the color the pigment appears to the eye.

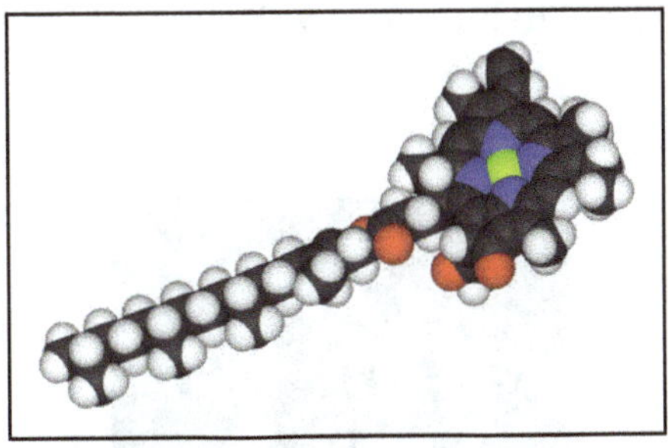

Space-filling model of the chlorophyll molecule.

Anthocyanin gives these pansies their dark purple pigmentation.

Chlorophyll is the primary pigment in plants; it is a porphyrin that absorbs red and blue wavelengths of light while reflecting green. It is the presence and relative abundance of chlorophyll that gives plants their green color. All land plants and green algae possess two forms of this pigment: chlorophyll *a* and chlorophyll *b*. Kelps, diatoms, and other photosynthetic heterokonts contain chlorophyll *c* instead of *b*, red algae possess chlorophyll *a* and " d". All chlorophylls serve as the primary means plants use to intercept light to fuel photosynthesis.

Carotenoids are red, orange, or yellow tetraterpenoids. They function as accessory pigments in plants, helping to fuel photosynthesis by gathering wavelengths of light not readily absorbed by chlorophyll. The most familiar carotenoids are carotene (an orange pigment found in carrots), lutein (a yellow pigment found in fruits and vegetables), and lycopene (the red pigment responsible for the color of tomatoes). Carotenoids have been shown to act as antioxidants and to promote healthy eyesight in humans.

Anthocyanins (literally "flower blue") are water-soluble flavonoid pigments that appear red to blue, according to pH. They occur in all tissues of higher plants, providing color in leaves, stems, roots, flowers, and fruits, though not always in sufficient quantities to be noticeable. Anthocyanins are most visible in the petals of flowers, where they may make up as much as 30% of the dry weight of the tissue. They are also responsible for the purple color seen on the underside of tropical shade plants such as *Tradescantia zebrina*. In these plants, the anthocyanin catches light that has passed through the leaf and reflects it back towards regions bearing chlorophyll, in order to maximize the use of available light.

Betalains are red or yellow pigments. Like anthocyanins they are water-soluble, but unlike anthocyanins they are indole-derived compounds synthesized from tyrosine. This class of pigments is found only in the Caryophyllales (including cactus and amaranth), and never co-occur in plants with anthocyanins. Betalains are responsible for the deep red color of beets, and are used commercially as food-coloring agents. Plant physiologists are uncertain of the function that betalains have in plants which possess them, but there is some preliminary evidence that they may have fungicidal properties.

Signals and Regulators

A mutation that stops *Arabidopsis thaliana* responding to auxin causes abnormal growth (right).

Plants produce hormones and other growth regulators which act to signal a physiological response in their tissues. They also produce compounds such as phytochrome that are sensitive to light and which serve to trigger growth or development in response to environmental signals.

Plant Hormones

Plant hormones, known as plant growth regulators (PGRs) or phytohormones, are chemicals that regulate a plant's growth. According to a standard animal definition, hormones are signal molecules produced at specific locations, that occur in very low concentrations, and cause altered processes in target cells at other locations. Unlike animals, plants lack specific hormone-producing tissues or organs. Plant hormones are often not transported to other parts of the plant and production is not limited to specific locations.

Plant hormones are chemicals that in small amounts promote and influence the growth, development and differentiation of cells and tissues. Hormones are vital to plant growth; affecting processes in plants from flowering to seed development, dormancy, and germination. They regulate which tissues grow upwards and which grow downwards, leaf formation and stem growth, fruit development and ripening, as well as leaf abscission and even plant death.

The most important plant hormones are abscissic acid (ABA), auxins, ethylene, gibberellins, and cytokinins, though there are many other substances that serve to regulate plant physiology.

Photomorphogenesis

While most people know that light is important for photosynthesis in plants, few realize that plant sensitivity to light plays a role in the control of plant structural development (morphogenesis). The use of light to control structural development is called photomorphogenesis, and is dependent upon the presence of specialized photoreceptors, which are chemical pigments capable of absorbing specific wavelengths of light.

Plants use four kinds of photoreceptors: phytochrome, cryptochrome, a UV-B photoreceptor, and protochlorophyllide *a*. The first two of these, phytochrome and cryptochrome, are photoreceptor proteins, complex molecular structures formed by joining a protein with a light-sensitive pigment. Cryptochrome is also known as the UV-A photoreceptor, because it absorbs ultraviolet light in the long wave "A" region. The UV-B receptor is one or more compounds not yet identified with certainty, though some evidence suggests carotene or riboflavin as candidates. Protochlorophyllide *a*, as its name suggests, is a chemical precursor of chlorophyll.

The most studied of the photoreceptors in plants is phytochrome. It is sensitive to light in the red and far-red region of the visible spectrum. Many flowering plants use it to regulate the time of flowering based on the length of day and night (photoperiodism) and to set circadian rhythms. It also regulates other responses including the germination of seeds, elongation of seedlings, the size, shape and number of leaves, the synthesis of chlorophyll, and the straightening of the epicotyl or hypocotyl hook of dicot seedlings.

Photoperiodism

Many flowering plants use the pigment phytochrome to sense seasonal changes in day length, which they take as signals to flower. This sensitivity to day length is termed photoperiodism. Broadly speaking, flowering plants can be classified as long day plants, short day plants, or day neutral plants, depending on their particular response to changes in day length. Long day plants require a certain minimum length of daylight to starts flowering, so these plants flower in the spring or summer. Conversely, short day plants flower when the length of daylight falls below a certain critical level. Day neutral plants do not initiate flowering based on photoperiodism, though some may use temperature sensitivity (vernalization) instead.

Although a short day plant cannot flower during the long days of summer, it is not actually the period of light exposure that limits flowering. Rather, a short day plant requires a minimal length of uninterrupted darkness in each 24-hour period (a short daylength) before floral development can begin. It has been determined experimentally that a short day plant (long night) does not flower if a flash of phytochrome activating light is used on the plant during the night.

The poinsettia is a short-day plant, requiring two months of long nights prior to blooming.

Plants make use of the phytochrome system to sense day length or photoperiod. This fact is utilized by florists and greenhouse gardeners to control and even induce flowering out of season, such as the *Poinsettia*.

Environmental Physiology

Paradoxically, the subdiscipline of environmental physiology is on the one hand a recent field of study in plant ecology and on the other hand one of the oldest. Environmental physiology is the preferred name of the subdiscipline among plant physiologists, but it goes by a number of other names in the applied sciences. It is roughly synonymous with ecophysiology, crop ecology, horticulture and agronomy. The particular name applied to the subdiscipline is specific to the viewpoint and goals of research. Whatever name is applied, it deals with the ways in which plants respond to their environment and so overlaps with the field of ecology.

Phototropism in *Arabidopsis thaliana* is regulated by blue to UV light.

Environmental physiologists examine plant response to physical factors such as radiation (including light and ultraviolet radiation), temperature, fire, and wind. Of particular importance are water relations (which can be measured with the Pressure bomb) and the stress of drought or inundation, exchange of gases with the atmosphere, as well as the cycling of nutrients such as nitrogen and carbon.

Environmental physiologists also examine plant response to biological factors. This includes not only negative interactions, such as competition, herbivory, disease and parasitism, but also positive interactions, such as mutualism and pollination.

Tropisms and Nastic Movements

Plants may respond both to directional and non-directional stimuli. A response to a directional stimulus, such as gravity or sunlight, is called a tropism. A response to a nondirectional stimulus, such as temperature or humidity, is a nastic movement.

Tropisms in plants are the result of differential cell growth, in which the cells on one side of the plant elongates more than those on the other side, causing the part to bend toward the side with less growth. Among the common tropisms seen in plants is phototropism, the bending of the plant toward a source of light. Phototropism allows the plant to maximize light exposure in plants which require additional light for photosynthesis, or to minimize it in plants subjected to intense light and heat. Geotropism allows the roots of a plant to determine the direction of gravity and grow downwards. Tropisms generally result from an interaction between the environment and production of one or more plant hormones.

Nastic movements results from differential cell growth (e.g. epinasty and hiponasty), or from changes in turgor pressure within plant tissues (e.g., nyctinasty), which may occur rapidly. A familiar example is thigmonasty (response to touch) in the Venus fly trap, a carnivorous plant. The traps consist of modified leaf blades which bear sensitive trigger hairs. When the hairs are touched by an insect or other animal, the leaf folds shut. This mechanism allows the plant to trap and digest small insects for additional nutrients. Although the trap is rapidly shut by changes in internal cell pressures, the leaf must grow slowly to reset for a second opportunity to trap insects.

Plant Disease

Powdery mildew on crop leaves.

Economically, one of the most important areas of research in environmental physiology is that of phytopathology, the study of diseases in plants and the manner in which plants resist or cope with infection. Plant are susceptible to the same kinds of disease organisms as animals, including viruses, bacteria, and fungi, as well as physical invasion by insects and roundworms.

Because the biology of plants differs with animals, their symptoms and responses are quite different. In some cases, a plant can simply shed infected leaves or flowers to prevent the spread of disease, in a process called abscission. Most animals do not have this option as a means of controlling disease. Plant diseases organisms themselves also differ from those causing disease in animals because plants cannot usually spread infection through casual physical contact. Plant pathogens tend to spread via spores or are carried by animal vectors.

One of the most important advances in the control of plant disease was the discovery of Bordeaux mixture in the nineteenth century. The mixture is the first known fungicide and is a combination of copper sulfate and lime. Application of the mixture served to inhibit the growth of downy mildew that threatened to seriously damage the French wine industry.

History

Early History

Sir Francis Bacon published one of the first plant physiology experiments in 1627 in the book, *Sylva Sylvarum*. Bacon grew several terrestrial plants, including a rose, in water and concluded that soil was only needed to keep the plant upright. Jan Baptist van Helmont published what is considered the first quantitative experiment in plant physiology in 1648. He grew a willow tree for five years in a pot containing 200 pounds of oven-dry soil. The soil lost just two ounces of dry weight and van Helmont concluded that plants get all their weight from water, not soil. In 1699, John Woodward published experiments on growth of spearmint in different sources of water. He found that plants grew much better in water with soil added than in distilled water.

Jan Baptist van Helmont.

Stephen Hales is considered the Father of Plant Physiology for the many experiments in the 1727 book; though Julius von Sachs unified the pieces of plant physiology and put them together as a discipline. His *Lehrbuch der Botanik* was the plant physiology bible of its time.

Researchers discovered in the 1800s that plants absorb essential mineral nutrients as inorganic ions in water. In natural conditions, soil acts as a mineral nutrient reservoir but the soil itself is not essential to plant growth. When the mineral nutrients in the soil are dissolved in water, plant roots absorb nutrients readily, soil is no longer required for the plant to thrive. This observation is the basis for hydroponics, the growing of plants in a water solution rather than soil, which has become a standard technique in biological research, teaching lab exercises, crop production and as a hobby.

Economic Applications

Food Production

In horticulture and agriculture along with food science, plant physiology is an important topic relating to fruits, vegetables, and other consumable parts of plants. Topics studied include: *climatic* requirements, fruit drop, nutrition, ripening, fruit set. The production of food crops also hinges on the study of plant physiology covering such topics as optimal planting and harvesting times and post harvest storage of plant products for human consumption and the production of secondary products like drugs and cosmetics.

Plant Pathology

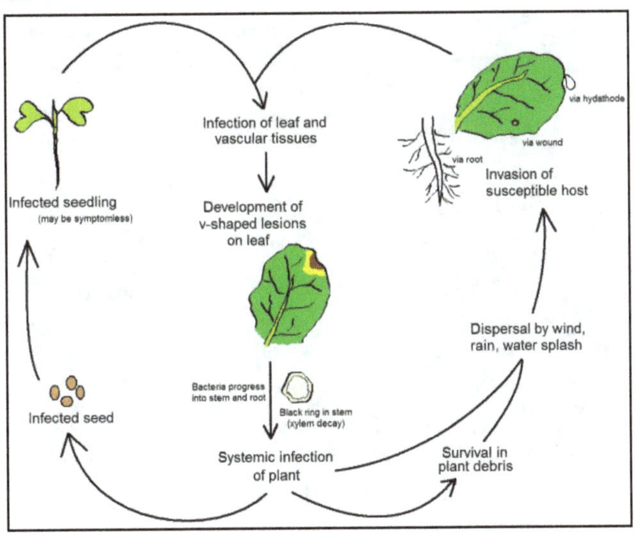

Life cycle of the black rot pathogen, *Xanthomonas campestris* pathovar *campes.*

Plant pathology (also phytopathology) is the scientific study of diseases in plants caused by pathogens (infectious organisms) and environmental conditions (physiological factors). Organisms that cause infectious disease include fungi, oomycetes, bacteria, viruses, viroids, virus-like organisms, phytoplasmas, protozoa, nematodes and parasitic plants. Not included are ectoparasites like insects, mites, vertebrate, or other pests that affect plant health by consumption of plant tissues. Plant pathology also involves the study of pathogen identification, disease etiology, disease cycles, economic impact, plant disease epidemiology, plant disease resistance, how plant diseases affect humans and animals, pathosystem genetics, and management of plant diseases.

Overview

Control of plant diseases is crucial to the reliable production of food, and it provides significant reductions in agricultural use of land, water, fuel and other inputs. Plants in both natural and cultivated populations carry inherent disease resistance, but there are numerous examples of devastating plant disease impacts, as well as recurrent severe plant diseases. However, disease control is reasonably successful for most crops. Disease control is achieved by use of plants that have been bred for good resistance to many diseases, and by plant cultivation approaches such as crop rotation, use of pathogen-free seed, appropriate planting date and plant density, control of field moisture, and pesticide use. Across large regions and many crop species, it is estimated that diseases typically reduce plant yields by 10% every year in more developed settings, but yield loss to diseases often exceeds 20% in less developed settings. Continuing advances in the science of plant pathology are needed to improve disease control, and to keep up with changes in disease pressure caused by the ongoing evolution and movement of plant pathogens and by changes in agricultural practices. Plant diseases cause major economic losses for farmers worldwide. The Food and Agriculture Organization estimates indeed that pests and diseases are responsible for about 25% of crop loss. To solve this issue, new methods are needed to detect diseases and pests early, such as novel sensors that detect plant odours and spectroscopy and biophotonics that are able to diagnos-tic plant health and metabolism.

Plant Pathogens

Powdery mildew, a biotrophic fungus.

Fungi

Most phytopathogenic fungi belong to the Ascomycetes and the Basidiomycetes.

The fungi reproduce both sexually and asexually via the production of spores and other structures. Spores may be spread long distances by air or water, or they may be soilborne. Many soil inhabiting fungi are capable of living saprotrophically, carrying out the part of their life cycle in the soil. These are known as facultative saprotrophs.

Fungal diseases may be controlled through the use of fungicides and other agriculture practices. However, new races of fungi often evolve that are resistant to various fungicides.

Rice blast, caused by a necrotrophic fungus.

Biotrophic fungal pathogens colonize living plant tissue and obtain nutrients from living host cells. Necrotrophic fungal pathogens infect and kill host tissue and extract nutrients from the dead host cells.

Significant fungal plant pathogens include:

Ascomycetes

- *Fusarium* spp. (causal agents of Fusarium wilt disease).
- *Thielaviopsis* spp. (causal agents of: canker rot, black root rot, *Thielaviopsis* root rot).
- *Verticillium* spp.
- *Magnaporthe grisea* (causal agent of rice blast).
- *Sclerotinia sclerotiorum* (causal agent of cottony rot).

Basidiomycetes

- *Ustilago* spp. (causal agents of smut).
- *Rhizoctonia* spp.
- *Phakospora pachyrhizi* (causal agent of soybean rust).
- *Puccinia* spp. (causal agents of severe rusts of virtually all cereal grains and cultivated grasses).
- *Armillaria* spp. (the so-called honey fungus species, which are virulent pathogens of trees and produce edible mushrooms).

Fungus-like Organisms

Oomycetes

The oomycetes are not true fungi but are fungus-like organisms. They include some of the most destructive plant pathogens including the genus *Phytophthora*, which includes the causal agents of potato late blight and sudden oak death. Particular species of oomycetes are responsible for root rot.

Despite not being closely related to the fungi, the oomycetes have developed very similar infection strategies. Oomycetes are capable of using effector proteins to turn off a plant's defenses in its infection process. Plant pathologists commonly group them with fungal pathogens.

Significant oomycete plant pathogens:

- *Pythium* spp.
- *Phytophthora* spp., including the causal agent of the Great Irish Famine (1845–1849).

Phytomyxea

Some slime molds in Phytomyxea cause important diseases, including club root in cabbage and its relatives and powdery scab in potatoes. These are caused by species of *Plasmodiophora* and *Spongospora*, respectively.

Bacteria

Most bacteria that are associated with plants are actually saprotrophic and do no harm to the plant itself. However, a small number, around 100 known species, are able to cause disease. Bacterial diseases are much more prevalent in subtropical and tropical regions of the world.

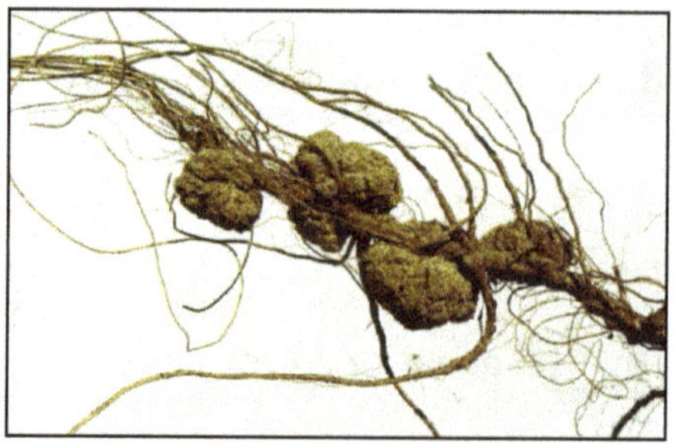

Crown gall disease caused by Agrobacterium.

Most plant pathogenic bacteria are rod-shaped (bacilli). In order to be able to colonize the plant they have specific pathogenicity factors. Five main types of bacterial pathogenicity factors are known: uses of cell wall–degrading enzymes, toxins, effector proteins, phytohormones and exopolysaccharides.

Pathogens such as *Erwinia* species use cell wall–degrading enzymes to cause soft rot. *Agrobacterium* species change the level of auxins to cause tumours with phytohormones. Exopolysaccharides are produced by bacteria and block xylem vessels, often leading to the death of the plant.

Bacteria control the production of pathogenicity factors via quorum sensing.

Vitis vinifera with "Ca. Phytoplasma vitis" infection.

Significant bacterial plant pathogens:

- Burkholderia.
- Proteobacteria.
    - *Xanthomonas* spp.
    - *Pseudomonas* spp.
- Pseudomonas syringae pv. tomato causes tomato plants to produce less fruit, and it "continues to adapt to the tomato by minimizing its recognition by the tomato immune system."

Phytoplasmas ('Mycoplasma-like Organisms') and Spiroplasmas

Phytoplasma and *Spiroplasma* are a genre of bacteria that lack cell walls and are related to the mycoplasmas, which are human pathogens. Together they are referred to as the mollicutes. They also tend to have smaller genomes than most other bacteria. They are normally transmitted by sap-sucking insects, being transferred into the plants phloem where it reproduces.

Tobacco mosaic virus.

Viruses, Viroids and Virus-like Organisms

There are many types of plant virus, and some are even asymptomatic. Under normal circumstances, plant viruses cause only a loss of crop yield. Therefore, it is not economically viable to try to control them, the exception being when they infect perennial species, such as fruit trees.

Most plant viruses have small, single-stranded RNA genomes. However some plant viruses also have double stranded RNA or single or double stranded DNA genomes. These genomes may encode only three or four proteins: a replicase, a coat protein, a movement protein, in order to allow cell to cell movement through plasmodesmata, and sometimes a protein that allows transmission by a vector. Plant viruses can have several more proteins and employ many different molecular translation methods.

Plant viruses are generally transmitted from plant to plant by a vector, but mechanical and seed transmission also occur. Vector transmission is often by an insect (for example, aphids), but some fungi, nematodes, and protozoa have been shown to be viral vectors. In many cases, the insect and virus are specific for virus transmission such as the beet leafhopper that transmits the curly top virus causing disease in several crop plants.

Nematodes

Nematodes are small, multicellular wormlike animals. Many live freely in the soil, but there are some species that parasitize plant roots. They are a problem in tropical and subtropical regions of the world, where they may infect crops. Potato cyst nematodes (*Globodera pallida* and *G. rostochiensis*) are widely distributed in Europe and North and South America and cause $300 million worth of damage in Europe every year. Root knot nematodes have quite a large host range, whereas cyst nematodes tend to be able to infect only a few species. Nematodes are able to cause radical changes in root cells in order to facilitate their lifestyle.

Root-knot nematode galls

Protozoa and Algae

There are a few examples of plant diseases caused by protozoa (e.g., *Phytomonas*, a kinetoplastid). They are transmitted as zoospores that are very durable, and may be able to survive in a resting state in the soil for many years. They have also been shown to transmit plant viruses.

When the motile zoospores come into contact with a root hair they produce a plasmodium and invade the roots.

Some colourless parasitic algae (e.g., *Cephaleuros*) also cause plant diseases.

Parasitic Plants

Parasitic plants such as mistletoe and dodder are included in the study of phytopathology. Dodder, for example, is used as a conduit either for the transmission of viruses or virus-like agents from a host plant to a plant that is not typically a host or for an agent that is not graft-transmissible.

Common Pathogenic Infection Methods

- Cell wall-degrading enzymes: These are used to break down the plant cell wall in order to release the nutrients inside.
- Toxins: These can be non-host-specific, which damage all plants, or host-specific, which cause damage only on a host plant.
- Effector proteins: These can be secreted into the extracellular environment or directly into the host cell, often via the Type three secretion system. Some effectors are known to suppress host defense processes. This can include: reducing the plants internal signaling mechanisms or reduction of phytochemicals production. Bacteria, fungus and oomycetes are known for this function.

Physiological Plant Disorders

Significant abiotic disorders can be caused by:

- *Natural*
- Drought
- Frost damage and breakage by snow and hail
- Flooding and poor drainage
- Nutrient deficiency
- Salt deposition and other soluble mineral excesses (e.g., gypsum)
- Wind (windburn and breakage by hurricanes and tornadoes)
- Lightning and wildfire (also often man-made)
- *Man-made* (arguably not abiotic, but usually regarded as such)
- Soil compaction
- Pollution of air, soil, or both
- Salt from winter road salt application or irrigation
- Herbicide over-application
- Poor education and training of people working with plants (e.g. lawnmower damage to trees)
- Vandalism

Orchid leaves with viral infections.

Epidemiology

The study of phytopathology epidemics.

Disease Resistance

Management

Quarantine:

A diseased patch of vegetation or individual plants can be isolated from other, healthy growth. Specimens may be destroyed or relocated into a greenhouse for treatment or study. Another option is to avoid the introduction of harmful nonnative organisms by controlling all human traffic and activity (e.g., AQIS), although legislation and enforcement are crucial in order to ensure lasting effectiveness.

Cultural:

Farming in some societies is kept on a small scale, tended by peoples whose culture includes farming traditions going back to ancient times. (An example of such traditions would be lifelong training in techniques of plot terracing, weather anticipation and response, fertilization, grafting, seed care, and dedicated gardening.) Plants that are intently monitored often benefit from not only active external protection but also a greater overall vigor. While primitive in the sense of being the most labor-intensive solution by far, where practical or necessary it is more than adequate.

Plant Resistance:

Sophisticated agricultural developments now allow growers to choose from among systematically cross-bred species to ensure the greatest hardiness in their crops, as suited for a particular region's pathological profile. Breeding practices have been perfected over centuries, but with the advent of genetic manipulation even finer control of a crop's immunity traits is possible. The engineering of food plants may be less rewarding, however, as higher output is frequently offset by popular suspicion and negative opinion about this "tampering" with nature.

Chemical:

Many natural and synthetic compounds can be employed to combat the above threats. This method works by directly eliminating disease-causing or-ganisms or curbing their spread; however, it has been shown to have too broad an effect, typically, to be good for the local ecosystem. From an economic standpoint, all but the sim-plest natural additives may disqualify a product from "organic" status, potentially reducing the value of the yield.

Biological:

Crop rotation may be an effective means to prevent a parasitic population from becoming well-established, as an organism affecting leaves would be starved when the leafy crop is replaced by a tuberous type, etc. Other means to undermine parasites without attacking them directly may exist.

Integrated:

The use of two or more of these methods in combination offers a higher chance of effectiveness.

References

- Michael A Dirr; Charles W Heuser, jr. (2006). "2". The Reference Manual of Woody Plant Propagation (Second ed.). Varsity Press Inc. pp. 26, 28, 29. ISBN 0942375092
- "Plasmopara viticola, the Cause of Downy Mildew of Grapes". The Origin of Plant Pathology and The Potato Famine, and Other Stories of Plant Diseases. Retrieved 03, April 2020
- Jackson RW (editor). (2009). Plant Pathogenic Bacteria: Genomics and Molecular Biology. Caister Academic Press. ISBN 978-1-904455-37-0
- Nicole Davis (September 9, 2009). "Genome of Irish potato famine pathogen decoded". Haas et al. Broad Institute of MIT and Harvard. Retrieved 11, May 2020
- Evert, Ray Franklin and Esau, Katherine (2006) Esau's Plant anatomy: meristems, cells, and tissues of the plant body - their structure, function and development Wiley, Hoboken, New Jersey, page xv, ISBN 0-471-73843-3
- "Fusarium oxysporum : The End of the Banana Industry?». The Origin of Plant Pathology and The Potato Famine, and Other Stories of Plant Diseases. Retrieved 12, March 2020
- Schulze, Ernst-Detlef; et al. (2005). Plant Ecology – (Section 1.10.1: Herbivory). Springer. Retrieved April 24, 2012. ISBN 3-540-20833-X
- "Scientists discover how deadly fungal microbes enter host cells". (VBI) at Virginia Tech affiliates. Physorg. July 22, 2010. Retrieved 23, August 2020
- Raven, P. H., R. F. Evert, & S. E. Eichhorn. Biology of Plants, 7th ed., page 9. (New York: W. H. Freeman, 2005). ISBN 0-7167-1007-2

Chapter 5

Animal Biology

The study of any life form that is mobile and exhibits a heterotrophic mode of nutrition comes under animal biology. Animals come in different shapes and sizes but display similar characteristics at the cellular level. This chapter covers in detail all the varied topics that come under it, like animal physiology, animal anatomy and animal behavior.

Animal Biology includes both invertebrate and vertebrate animals and addresses how they work (form, function, behavior, physiology, ecology), how they develop (gene expression, cell biology, morphogenesis), and how they evolve (genetics, diversity, phylogeny). Animal biology provides a comprehensive overview of how the different levels of biology relate to one another, from molecules to behaviour, from physiology to ecology.

Animals are the most advanced organisms on the planet. The first animals are thought to have evolved over 550 million years ago and they have since spread to all corners of the globe and diversified into millions of species. Animals can be distinguished from other living things by the combination of three features: 1) they are unable to make their own food; 2) they have multiple cells; and each cell has a nucleus and other specialized structures called organelles. Scientifically put, animals are heterotrophic, multicellular organisms with eukaryotic cells.

Animal species are often split into vertebrates and invertebrates – two very broad categories of animals. Vertebrate species are animals that do have a backbone such as humans, birds, fish and reptiles and make up the majority of large land animals. Invertebrate animals lack a backbone and often have a skeleton on the exterior of the body. The majority of animal species are classed as invertebrates and many you will be familiar with such as insects, jellyfish, worms and spiders.

Mammals are a group of warm-blooded vertebrates and include the largest and smartest animals on the planet. They are distinguished from other animals by having hair or fur and mammary glands for milk production.

Reptiles are a group of animals that include snakes, lizards, crocodiles, turtles and tuatara. They are an ancient group of cold-blooded, egg laying vertebrates with scales or scutes rather than fur or feathers.

Birds are a class of animals known as Aves and are easily distinguished by their feathers and wings. They have an incredible ability to fly and often perform amazing migrations over hundreds and even thousands of miles.

Fish are a group of aquatic animals with skulls, gills and digit-less limbs. Fish were the first vertebrate animals to evolve and there is now more than 30,000 species living on Earth. The highest diversity of fish is found in tropical coral reefs.

An invertebrate is any animal without an internal backbone such as insects, shellfish, worms, and spiders. Invertebrates are incredibly diverse and make up roughly 95% of all animal species.

Characteristics of Animals

An animal pertains to any of the eukaryotic multicellular organisms that comprise the biological kingdom of Animalia. Animals possess several characteristics that set them apart from other living things. Some of their general characteristics are as follows:

- Animals are multicellular organisms: The animal body is composed of several cells performing specific functions as opposed to bacteria and most protists that are unicellular. The cells may then be organized into various animal tissues, such as epithelial tissues, connective tissues, muscle tissues, nervous tissues, and vascular tissues. The cells in tissues may be held through cell junctions, e.g. tight junctions, gap junctions, and desmosomes.
- Animals are eukaryotic: An animal cell contains a membrane-bound nucleus. The nucleus is the organelle that contains chromosomes that bear genes. Apart from the nucleus, there are other organelles suspended in the cytoplasm of an animal cell, such as the Golgi apparatus, endoplasmic reticulum, lysosomes, and peroxisomes. Animal cells lack plastids and cell walls, which are abundant in plant cells, algae, and certain fungi.
- Animals are heterotrophic: Unlike plants and algae that are autotrophic, the animals depend on another organism for food. Animal carnivores, for instance, are adapted to hunting their prey while others scavenge for animal carcasses or remains. Others feed on plants and as such are referred to as herbivores. Omnivores are animals that feed on both plants and animals. Most animals have an elaborate digestive system in the form of an internal chamber (a digestive tract) that processes ingested food to extract nutrients from it and then released from the body in the form of excreta or a waste matter. Many animals have a mouth for ingestion and anus for excretion. Other animals (e.g. Platyhelminthes, cnidarians) have only one opening that serves both as a mouth and an anus. Poriferans, in contrast, lack a digestive system (as well as nervous and circulatory systems).
- Animals are generally motile: They have the capacity to move at will. They can spontaneously and actively move by metabolically utilizing energy (e.g. ATP) during the process and with the aid of muscles and locomotory structures (e.g. arms, legs, wings, fins, tails, etc.). Animal locomotion refers to the variety of movements that animals use to move from one location to another. Some of these movements are running, walking, jumping, hopping, slithering, swimming, gliding, flying, soaring, and so on. Animals move for multifarious reasons. Some of the reasons animals move are to hunt prey, escape predators, and find a mate or a suitable habitat. There are animals, though, that have become sessile later in life. They become permanently attached to a substrate. Examples are barnacles, sponges, mussels, and corals.
- Animals possess specialized sensory organs such as eyes, ears, nose, skin, and tongue. These sensory organs are vital for use in recognizing and responding to stimuli in the environment. Each of these sense organs contains common and specialized receptors.
- Animals reproduce sexually: They produce a haploid sperm cell (a male sex cell) and a haploid ovum (a female sex cell) that unite at fertilization to form a diploid zygote. Conversely, some animals are capable of asexual reproduction. For instance, some cnidarians produce a genetic clone by budding. Others (e.g. aphids) are capable of parthenogenesis whereby they produce fertile eggs without mating.

- Animals develop from an embryo that passes through a blastula stage. In the early embryonic development, a fluid-filled cavity called blastocoel appears within a morula (i.e. a ball of cells that forms from a series of cell divisions). At this point, the embryo is referred to as a blastula. Cell movement and differentiation start to occur during this stage.
- Animals have a distinctive cell division when a cleavage furrow forms to separate daughter cells. This is in contrast to the plant's cell plate (phragmoplast) that forms in cell division.
- Animals respire aerobically, taking in oxygen (inspiration) and then releasing carbon dioxide (expiration). Oxygen is important to cell respiration as it serves as the final electron acceptor in redox reactions during the synthesis of metabolic energy. The different animal structures involved in the exchange of respiratory gases: (1) skin of tapeworms, earthworms, and leeches, (2) trachea of insects, (3) gills of fish, and (4) lungs of mammals, reptiles, and birds. Amphibians use different respiratory organs at different stages, i.e. gills at tadpole stage and then skin and lungs at the adult stage.
- With few exceptions, animals possess the following general biological systems: integumentary system, lymphatic system, muscular system, nervous system, reproductive system, respiratory system, skeletal system, and urinary system.

Animal Body Plan

Most animals (at least 99%) have a bilaterally symmetric body plan. When the body is divided by a sagittal plane, the result is having two sides with roughly mirror images, at least morphologically. Most animals are sexually dimorphic, i.e. males and females exhibit different characteristics aside from the differences in their sex organs. For example, males are typically taller and bigger in size than females. In another example, the plumage of birds and the scales of fish are usually more vibrant or colorful in males than in females, which seem to be essential in attracting mates.

In general, animals have the following body systems (as mentioned previously) and each of these systems is made up of organs and tissues. The organs, in turn, are made up of tissues that carry out a particular function. The tissues consist of cells that perform various metabolic activities. There are two major types of cells: somatic cells and sex cells (i.e. sperm cell and egg cell).

At the cellular level, an animal cell has compartmentalized structures called organelles. The different organelles are mitochondria, Golgi apparatus, endoplasmic reticulum, lysosomes, and peroxisomes. Plastids, which are abundant in a plant or an autotrophic cell, are not present in animal cells. The cell wall is also lacking. Thus, an animal cell in a hypotonic solution will swell and eventually burst (lyse) since it lacks the cell wall that prevents lysis during excessive osmosis. The nucleus contains chromosomes that can be classified into two types: autosomes and sex chromosomes.

Life Cycle

The life cycles in the animal kingdom are diverse. For brevity, the four fundamental stages are infancy, young, adulthood, senescence. All animals undergo these stages in their life cycle. Most mammals have a life cycle that is as simple and straightforward. Nevertheless, there are animals whose life cycles are rather complex. Amphibians, like frogs, undergo metamorphosis. Initially, tadpoles live underwater and breathe through the gills. Over time, they grow into froglets, and soon into adults that live on land and breathe through the lungs. Another example is insects. A gravid female releases eggs. Larva (young) hatches from the egg and goes on to feed. Soon, the larva becomes a pupa, which in turn, metamorphoses into an adult capable of breeding. These animals go through radical transformations. The adult is morphologically far-off from the young.

In animals, the major form of reproduction is by sexual means. Male and female sex cells unite to form a zygote that in time will grow into an animal of their kind. Most animals are diploid. This means that the progeny comes from the union of two sex cells and then grows possessing two sets of chromosomes – one from the male parent and the other from the female parent. Sex determination is usually genetic. It can be identified based on chromosomal combinations. The different chromosomal systems that determine the sex of the animal are XY, ZW, XO, ZO, UV, and haplodiploidy. In most mammals (including humans), sex determination is based on the XY sex-determination system wherein a female has two X chromosomes and a male has one X chromosome and one Y chromosome.

Some animals are capable of asexual reproduction, e.g. by fragmentation and parthenogenesis. In this case, the animal reproduces a clone. However, asexual reproduction could become an evolutionary disadvantage as it could lead to decreased diversity in the gene pool and the prevalence of harmful recessive traits.

Ecology

Animals are an important part of the ecological system. They form intricate food webs. They interact with other organisms and form various symbiotic relationships, e.g. mutualism, predation, competition, commensalism, and parasitism. Some animals demonstrate altruistic (self-sacrificing) behavior for the benefit of their conspecifics or colonies. The animals in the wild help regulate biodiversity as they act as nature's biological control agents as well as serve as an important source of nutrients for other species. Remains from dead animals serve as food for detritivores. Moreover, the decomposition of the dead animal body enables the return of elements and organic compounds to Earth. Thus, the extinction of an animal species causes a major impact on the ecosystem.

In 2011, the Census of Marine Life estimated that there could be around 8.74 million eukaryote species on Earth. Approximately, 7.77 million of them are animal species, and 953, 434 have already been described and cataloged. Unfortunately, there are possibilities that some of the animal species could have become extinct even before they are fully identified and cataloged. Thus, conservation efforts are being made globally to save the dwindling number of animals, especially those nearing extinction. The International Union for Conservation of Nature's Red List of Threatened Species, a system of assessing the conservation status of species worldwide, utilized a system of labeling species based on extinction risk. Accordingly, animal species are categorized as: "data deficient", "least concern", "near-threatened", "vulnerable", "endangered", "critically endangered", "regionally extinct", "extinct in the wild", and "extinct".

Evolution of Animals

A brief timeline of the evolution of animals is shown below:

- Proterozoic eon - Neoproterozoic era - Ediacaran period: Around 600 million years ago, multicellular organisms, referred to as Ediacaran biota, existed based on the first fossils recovered in South Australia. However, the biota disappeared after it flourished for the last 40 million years before the start of the Cambrian. Thus, they are also referred to as the

"Pre-Cambrian life". Called the Ediacaran fauna, these organisms are presumed as the earliest animal forms that resembled the sponges with several morphologies: mud-filled bags (e.g. Pteridinium), flattened torus (e.g. Vendoglossa tuberculata), "quilted" (e.g. Charnia, Swartpuntia), sturdy with spicules (e.g. Coronacollina acula). Size ranges from 1cm to less than 1m.

- Phanerozoic eon - Paleozoic era - Cambrian period: approximately 541 million years ago, several animals first appeared in the fossil record. Referred to as the Cambrian explosion, it was as if a sudden burst of life occurred during the Cambrian period. Fossils of diverse animals were recovered. One of the most popular Cambrian fossil material is the Burgess Shale. The fossil assemblage is a 508 million-year-old rock unit that contains diverse animals, such as trilobites, mollusks, echinoderms, brachiopods, Opabinia, Pikaia. Fossil of agnathan (jawless fish) Haikouichthys ercaicunensis was also found and hinted that the agnathans may have been the earliest fish and existed as early as 530 million years ago.

Animals are believed to have evolved in the sea, and animal evolution may have started billion years ago or at least long before the Ediacaran period. Eventually, animals ventured onto land, presumably during the Late Cambrian or Early Ordovician. The discovery of footprint of centipede-like, lobster-sized animal preserved in rocks in southeastern Canada hinted that animals might have ventured ashore about 530 million years ago probably to mate and lay eggs.

- Phanerozoic eon - Paleozoic era - Ordovician period: In the Ordovician period, the animals continued to diversify. The fauna was predominated by invertebrates such as trilobites, snails, shellfish, cephalopods, and corals. Primitive fish were also evolving during this period.3 However, a glaciation event led to the Ordovician-Silurian extinction (one of the deadliest mass extinctions in the history of the earth).

- Phanerozoic eon - Paleozoic era - Silurian period: In the Silurian period, a mass evolution of fish occurred. The Silurian marine fauna was predominated by jawless (agnathans) and jawed fish. The first freshwater fish first appeared as well. The arthropods, though, remained as the major predators.

Terrestrial animals such as early arachnids and arthropods (together with vascular plants) not just ventured the land. This time, they colonized it for the first time. Thus, land animals in due course evolved an internal gas exchange systems, waterproof external layers, skeletal systems (endo- or exoskeletons), and a form of reproduction that does not involve water.

- Phanerozoic eon - Paleozoic era - Devonian period: Devonian period is referred to as the Age of the Fish. Sharks and agnathans became common during this period. Two lineages, ray-finned and lobe-finned fish, first appeared in fossil records. The fish became the dominant marine vertebrate species.

On land, the appearance of primitive plants, trees, and shrub-like forests provided new habitats for arthropods. The first tetrapods to appear were amphibians (e.g. Ichthyostega). The land continued to be dominated by arthropods, such as arachnids and wingless insects. However, another major mass extinction occurred. Called as the late Devonian extinction, about 70% of species became extinct.

- Phanerozoic eon - Paleozoic era - Carboniferous period: In the Carboniferous period, some of the dominant invertebrates in marine habitats are foraminifera, corals, bryozoan, ostracods, brachiopods, and echinoderms. In freshwater, the dominant invertebrates are bivalve mollusks and crustaceans. As for the dominant vertebrates, fish dominated the aquatic habitats. Major predators include elasmobranchs in the seas and Xenacanthida in freshwaters.

Arthropods on land continued to evolve. Some of them grew into giants. For example, the millipede-likeArthropleura is regarded as the largest land invertebrate as it could grow to size up to 8.5 ft. The dragonfly-like insect Meganeura, in turn, is the largest flying insect.

One of the most important evolutionary milestones was the appearance of tetrapods that lay amniotic eggs. The laying of amniotic eggs in a drier environment allowed tetrapod ancestors of reptiles, birds, and mammals to move farther away from the waterside, and thereby dominate farther inland. Because of this, amniotes diversified greatly by the end of this period.

- Phanerozoic eon - Paleozoic era - Permian period: The Permian period had one continent called Pangaea surrounded by Panthalassa ocean, and as such the inlands were very dry and arid. Thus, reptiles and synapsids were the animals that flourished during this period. A major evolutionary transition occurred as the reptile group "Dimetrodon" evolved and gave rise to "beast-faced" therapsids. These reptiles evolved and gave rise to the cynodonts (early ancestors of mammals) towards the end of the period. The first archosaurs (early ancestors of dinosaurs) also appeared in late Permian period.

Another mass extinction event called "the Great Dying" occurred and caused the death of about 90% of life on Earth.

- Phanerozoic eon - Mesozoic era: During this era, the reptiles dominated the land, the seas, and the air. It is referred to as "the Age of the dinosaurs". Towards the end of the Cretaceous period, a mass extinction occurred and caused the wiping out of the dinosaurs and other large animals (i.e. >25 kg in weight). This led to the expansion of the mammals on land.

- Phanerozoic eon - Cenozoic era: This era called the "new life" is the most recent geological era that spans from 66 million years ago to the present day. The era features the rise of mammals. The great apes evolved and led to the evolution of hominids, which was the evolutionary line that led to the Homo species. The only extant species of the genus Homo is the Homo sapiens (anatomically modern humans).

Taxonomy

In 1758, Carl Linnaeus created a biological classification system for animals. In 1874, Ernst Haeckel proposed to divide the animal kingdoms into two groups: the multicellular Metazoa and the single-celled Protozoa. Later, the protozoans are no longer considered as animals and have been designated with their own biological kingdom. At present, the kingdom Animalia is comprised of the following extant phyla:

- Acanthocephala (thorny-headed worms),

- Acoelomorpha (acoels),
- Annelida (segmented worms),
- Arthropoda (arthropods),
- Brachiopoda (lamp shells),
- Bryozoa (moss animals, sea mats),
- Chaetognatha (arrow worms),
- Chordata (chordates),
- Cnidaria (coelenterates),
- Ctenophora (comb jellies),
- Echinodermata (echinoderms),
- Entoprocta (goblet worm),
- Gastrotricha (meiofauna),
- Gnathostomulida (jaw worms),
- Hemichordata (acorn worms, pterobranchs),
- Kinorhyncha (mud dragons),
- Loricifera (brush heads),
- Micrognathozoa (tiny jaw animals),
- Mollusca (molluscs),
- Nematoda (nematodes, roundworms),
- Nematomorpha (horsehair worms),
- Nemertea (ribbon worms),
- Onychophora (velvet worms),
- Orthonctida,
- Phoronida (horseshoe worms),
- Placozoa (plate animals),
- Platyhelminthes (flatworms),
- Porifera (sponges),
- Priapulida,

- Rhombozoa,
- Rotifera (rotifers),
- Sipuncula (peanut worms),
- Tardigrada (water bears),
- Xenoturbellida.

Significance

Animals have long been a key player in the ecosystem and performing an essential role in the lives of other organisms whether from the same kingdom or not. For instance, carbon dioxide that the animals breathe out is essential to plants and other autotrophs as it is a reactant in the early steps of photosynthesis.

Animals have also a great impact on the lives of humans. Besides as a source of meat, milk, and eggs, animals are also a source of materials (e.g. wool). Because of this, some of them have been domesticated. Cows, chicken, horses, pigs, sheep, etc. are some of the domesticated animals. There are also animals that are made as pets, e.g. dogs, cats, birds, etc. Some of them are even trained to carry out "jobs", e.g. guard dogs, rodent-hunting cats, homing pigeons. Certain animals (e.g. fruit flies, mice, primates) are made to serve as animal models for scientific research.

Trivia

- The largest extant animal on earth is the blue whale (Balaenoptera musculus). It could weigh up to 190 metric tonnes and reach up to 110 ft long.
- On land, the largest extant animal is the African bush elephant (Loxodonta africana) that could weigh up to 12.25 tonnes and reach up to 35.0 ft long.
- The smallest animal is a cnidarian Myxozoa (e.g. Myxobolus shekel) that grows no larger than 20 μm.

Characteristics of the Animal Body

Every animal has a distinct body plan, adapted in response to environmental pressures, that limits its size and shape.

Animal Form and Function

Animals vary in form and function. From a sponge to a worm to a goat, an organism has a distinct body plan that limits its size and shape. The term body plan is the "blueprint" encompassing aspects such as symmetry, segmentation, and limb disposition. Body plans have been considered to have evolved in a geologically-sudden flash during the Cambrian Explosion (roughly 542 million years ago). However, there is also evidence of a more gradual development of body plans. With a few exceptions, most notably the sponges and Placozoa, animals have bodies differentiated into separate tissues, which in turn make up more complex organs and organ systems. These include tissues such as muscles, which are able to contract and control locomotion, and nerves, which send and process signals. Typically, there is also an internal digestive chamber with one or two openings. Animals' bodies are also designed to interact with their environments, whether in the deep sea, a rainforest canopy, or the desert. In addition, animal body plans have evolved in response to environmental pressures, as observed in fossil records, in order to enhance survival and reproductive success. Therefore, a large amount of information about the structure of an organism's body (anatomy) and the function of its cells, tissues, and organs (physiology) can be learned by studying that organism's environment. The arctic fox is an example of a complex animal that has adapted to its environment and illustrates the relationships between an animal's form and function.

Limits on Animal Size and Shape

Animal shape and body size are influenced by environmental factors as well as the presence of an exoskeleton or an endoskeleton. Animals with bilateral symmetry that live in water tend to have a fusiform shape: a tubular shaped body that is tapered at both ends. This shape decreases the drag on the body as it moves through water and allows the animal to swim at high speeds. Certain types of sharks can swim at fifty kilometers an hour, while some dolphins can swim at 32-40 kilometers per hour. Land animals frequently travel faster (although the tortoise and snail are significantly slower than sharks or dolphins). Another difference in the adaptations of aquatic and land-dwelling organisms is that aquatic organisms are constrained in shape by the forces of drag in the water since water has higher viscosity than air. However, land-dwelling organisms are constrained mainly by gravity; drag is relatively unimportant. For example, most adaptations in birds are for gravity, not for drag.

Maximum Speed of Assorted Land Marine Animals		
Animal	**Speed (kmh)**	**Speed (mph)**
Cheetah	113	70
Quarter horse	77	48
Fox	68	42
Shortfin mako shark	50	31
Domestic house cat	48	30
Human	45	28
Dolphin	32–40	20–25
Mouse	13	8
Snail	0.05	0.03

Animal speeds: Land and marine animals travel at varying speeds. Land animals usually travel at higher speeds, but marine animals such as dolphins and sharks travel relatively fast.

Most animals have an exoskeleton, including insects, spiders, scorpions, horseshoe crabs, centipedes, and crustaceans. Scientists estimate that, of insects alone, there are over 30 million species on our planet. The exoskeleton is a hard covering or shell that provides benefits to the animal, such as protection against damage from predators and from water loss (for land animals); it also provides for the attachments of muscles. As the tough and resistant outer cover of an arthropod, the exoskeleton may be constructed of a tough polymer, such as chitin, and is often biomineralized with materials, such as calcium carbonate. This is fused to the animal's epidermis. Ingrowths of the exoskeleton called apodemes function as attachment sites for muscles, similar to tendons in more advanced animals. In order to grow, the animal must first synthesize a new exoskeleton underneath the old one and then shed or molt the original covering. This limits the animal's ability to grow continually. It may limit the individual's ability to mature if molting does not occur at the proper time. The thickness of the exoskeleton must be increased significantly to accommodate any increase in weight. It is estimated that a doubling of body size increases body weight by a factor of eight. The increasing thickness of the chitin necessary to support this weight limits most animals with an exoskeleton to a relatively-small size.

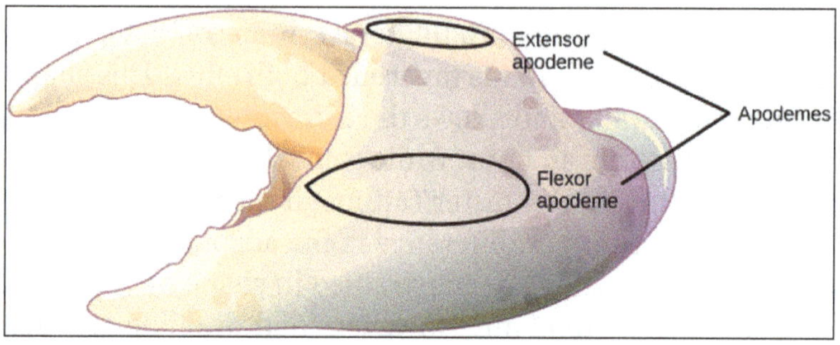

Apodemes: Apodemes are ingrowths on arthropod exoskeletons to which muscles attach. The apodemes on this crab leg are located above and below the fulcrum of the claw. Contraction of muscles attached to the apodemes pulls the claw closed.

The same principles apply to endoskeletons, but they are more efficient because muscles are attached on the outside, making it easier to compensate for increased mass. An animal with an endoskeleton has its size determined by the amount of skeletal system it needs in order to support the other tissues and the amount of muscle it needs for movement. As the body size increases, both bone and muscle mass increase. The speed achievable by the animal is a balance between its overall size and the bone and muscle that provide support and movement.

Limiting Effect of Diffusion on Size and Development

Less efficient diffusion in larger cells led to multicellular organisms with specialized tissues that supply nutrients and remove waste. The exchange of nutrients and wastes between a cell and its watery environment occurs through the process of diffusion. All living cells are bathed in liquid, whether they are in a single-celled organism or a multicellular one. Diffusion is effective over a specific distance and limits the size that an individual cell can attain. If a cell is a single-celled microorganism, such as an amoeba, it can satisfy all of its nutrient and waste needs through diffusion. If the cell is too large, then diffusion is ineffective at completing all of these tasks. The center of the cell does not receive adequate nutrients nor is it able to effectively dispel its waste.

An important concept in understanding the efficiency of diffusion as a transportation mechanism is the surface-to-volume ratio. Recall that any three-dimensional object has a surface area and volume; the ratio of these two quantities is the surface-to-volume ratio. Consider a cell shaped like a perfect sphere: it has a surface area of $4\pi r^2$, and a volume of $(4/3)\pi r^3$. The surface-to-volume ratio of a sphere is 3/r; as the cell gets bigger, its surface-to-volume ratio decreases, and making diffusion less efficient. The larger the size of the sphere, or animal, the less surface area for diffusion it possesses.

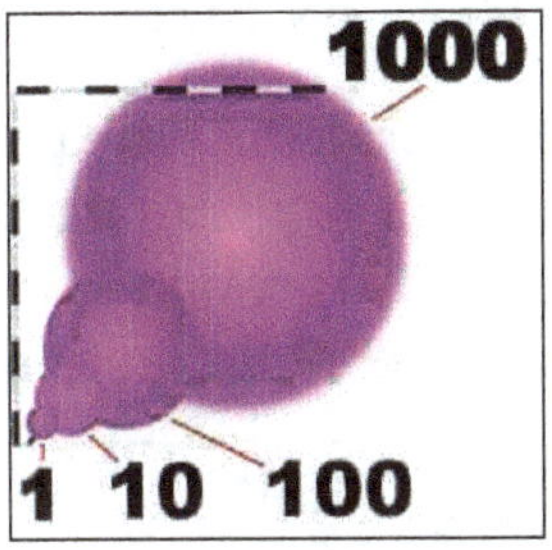

Surface-to-volume ratio: The image illustrates the comparison of spheres of one to one thousand volume units. The surface-to-volume ratio of a sphere decreases as the sphere gets bigger. The surface area of a sphere is $4\pi r^2$ and it has a volume of $(4/3)\pi r^3$ which makes the surface-to-volume ratio 3/r. This has an effect on diffusion because it relies on the surface area of a cell: as a cell gets bigger, diffusion becomes less efficient.

The solution to producing larger organisms is for them to become multicellular. Specialization occurs in complex organisms, allowing cells to become less efficient at completing all tasks since they are now more efficient at doing fewer tasks. For example, circulatory systems bring nutrients and remove waste, while respiratory systems provide oxygen for the cells and remove carbon dioxide from them. Other organ systems have developed further specialization of cells and tissues and efficiently control body functions. Surface-to-volume ratio also applies to other areas of animal development, such as the relationship between muscle mass and cross-sectional surface area in supporting skeletons or in the relationship between muscle mass and the generation of dissipation of heat.

Animal Bioenergetics

An animal's body size, activity level, and environment impacts the ways it uses and obtains energy. All animals must obtain their energy from food they ingest or absorb. These nutrients are converted to adenosine triphosphate (ATP) for short-term storage and use by all cells. Some animals store energy for slightly longer times as glycogen, while others store energy for much longer times in the form of triglycerides housed in specialized adipose tissues. No energy system is one hundred percent efficient as an animal's metabolism produces waste energy in the form of heat. If an animal can conserve that heat and maintain a relatively-constant body temperature, it is classified as a warm-blooded animal: an endotherm. The insulation used to conserve the body heat comes in the forms of fur, fat, or feathers. The absence of insulation in ectothermic animals increases their dependence on the environment for body heat.

The amount of energy expended by an animal over a specific time is called its metabolic rate. The rate is measured in joules, calories, or kilocalories (1000 calories). Carbohydrates and proteins contain about 4.5-5 kcal/g, while fat contains about 9 kcal/g. Metabolic rate is estimated as the basal metabolic rate (BMR) in endothermic animals at rest and as the standard metabolic rate (SMR) in ectotherms. Human males have a BMR of 1600-1800 kcal/day, and human females have a BMR of 1300 to 1500 kcal/day. Even with insulation, endothermal animals require extensive amounts of energy to maintain a constant body temperature. An ectotherm such as an alligator has an SMR of 60 kcal/day.

Energy Requirements Related to Body Size

Smaller endothermic animals have a greater surface area for their mass than larger ones. Therefore, smaller animals lose heat at a faster rate than larger animals and require more energy to maintain a constant internal temperature. This results in a smaller endothermic animal having a higher BMR, per body weight, than a larger endothermic animal.

Species		
Mass	35 g	4,500,000 g
Metabolic rate	890 mm^3 0_2/g body mass/hr	75 mm^3 0_2/g body mass/hr

Body size and metabolic rate: The mouse has a much higher metabolic rate than the elephant since it has greater surface area relative to mass.

Energy Requirements Related to Levels of Activity

The more active an animal is, the more energy is needed to maintain that activity and the higher its BMR or SMR. The average daily rate of energy consumption is about two to four times an animal's BMR or SMR. Humans are more sedentary than most animals and have an average daily rate of only 1.5 times the BMR. The diet of an endothermic animal is determined by its BMR.

Energy Requirements Related to Environment

Animals adapt to extremes of temperature or food availability through torpor. Torpor is a process that leads to a decrease in activity and metabolism, which allows animals to survive adverse conditions. Torpor can be used by animals for long periods. For example, animals can enter a state of hibernation during the winter months, which enables them to maintain a reduced body temperature. During hibernation, ground squirrels can achieve an abdominal temperature of 0 °C (32 °F), while a bear's internal temperature is maintained higher at about 37 °C (99 °F).

If torpor occurs during the summer months with high temperatures and little water, it is called estivation. Some desert animals estivate to survive the harshest months of the year. Torpor can occur on a daily basis; this is seen in bats and hummingbirds. While endotherm is limited in smaller animals by surface-to-volume ratio, some organisms can be smaller and still be endotherms because they employ daily torpor during the part of the day that is coldest. This allows them to conserve energy during the colder parts of the day when they consume more energy to maintain their body temperature.

Animal Body Planes and Cavities

Vertebrates can be divided along different planes in order to reference the locations of defined cavities. A standing vertebrate animal can be divided by several planes that can be used to as references to describe locations of body parts or organs. A sagittal plane divides the body into right and left portions. A midsagittal plane divides the body exactly in the middle, making two equal right and left halves. A frontal plane (also called a coronal plane) separates the front (ventral) from the back (dorsal). A transverse plane (or, horizontal plane) divides the animal into upper and lower portions. This is sometimes called a cross section; if the transverse cut is at an angle, it is called an oblique plane.

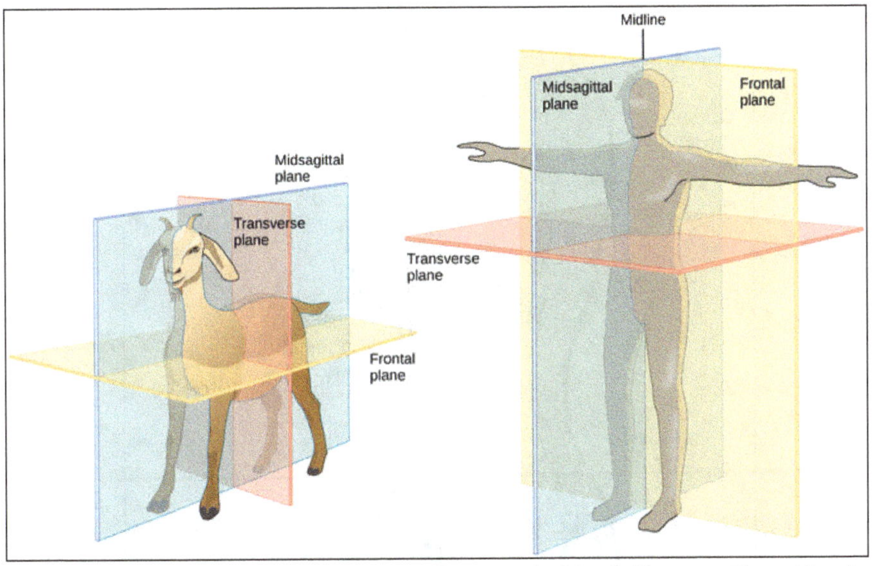

Body planes: Shown are the planes of a quadruped goat and a bipedal human. The midsagittal plane divides the body exactly in half into right and left portions. The frontal plane divides the front and back, while the transverse plane divides the body into upper and lower portions.

Vertebrate animals have a number of defined body cavities. The posterior (dorsal) and anterior (ventral) cavities are each subdivided into smaller cavities. In the posterior cavity, the cranial cavity houses the brain and the spinal cavity (or vertebral cavity) encloses the spinal cord. Just as the brain and spinal cord make up a continuous, uninterrupted structure, the cranial and spinal cavities that house them are also continuous. The brain and spinal cord are protected by the bones of the skull and vertebral column and by cerebrospinal fluid, a colorless fluid produced by the brain, which cushions the brain and spinal cord within the posterior (dorsal) cavity.

Body cavities: Vertebrate animals have two major body cavities. The dorsal cavity, indicated in green, contains the cranial and the spinal cavity. The ventral cavity, indicated in yellow, contains the thoracic cavity and the abdominopelvic cavity. The thoracic cavity is separated from the abdominopelvic cavity by the diaphragm. The abdominopelvic cavity is separated into the abdominal cavity and the pelvic cavity by an imaginary line parallel to the pelvis bones.

The anterior cavity has two main subdivisions: the thoracic cavity and the abdominopelvic cavity. The thoracic cavity is the more superior subdivision of the anterior cavity and is enclosed by the rib cage. The thoracic cavity contains the pleural cavity around the lungs and the pericardial cavity, which surrounds the heart. The diaphragm forms the floor of the thoracic cavity, separating it from the more inferior abdominopelvic cavity. The abdominopelvic cavity is the largest cavity in the body. Although no membrane physically divides the abdominopelvic cavity, it can be useful to distinguish between the abdominal cavities, the division that houses the digestive organs from the pelvic cavity, the division that houses the organs of reproduction.

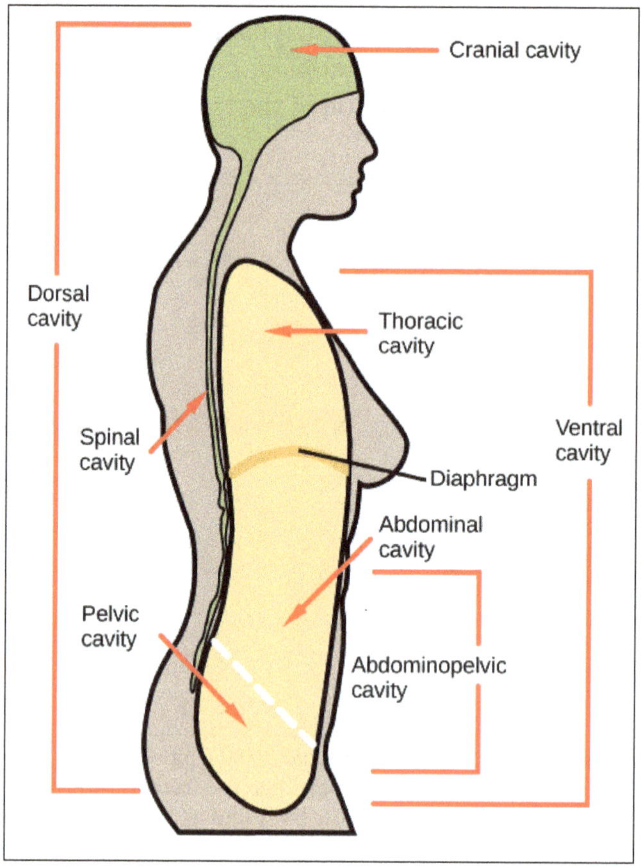

Animal Physiology

Animal physiology is the study of how animals work, and investigates the biological processes that occur for animal life to exist. These processes can be studied at various levels of organization from membranes through to organelles, cells, organs, organ systems, and to the whole animal. Animal physiology examines how biological processes function, how they operate under various environmental conditions, and how these processes are regulated and integrated. The study of animal physiology is closely linked with anatomy (i.e., the relationship of function with structure) and with the basic physical and chemical laws that constrain living as well as nonliving systems. Although all animals must function within basic physical and chemical constraints, there is a diversity of mechanisms and processes by which different animals work. A comparative approach to animal physiology highlights underlying principles, and reveals diverse solutions to various environmental challenges. It can reveal similar solutions to a common problem, or modifications of a particular physiological system to function under diverse conditions.

Circulatory System

The circulatory systems are a network of blood vessels supplying the body with oxygen and nutrients, while removing carbon dioxide and waste. Most animals are complex, multicellular organisms that require a mechanism for transporting nutrients throughout their bodies and for removing waste products. The circulatory system has evolved over time from simple diffusion through cells, in the early evolution of animals, to a complex network of blood vessels that reach all parts of the human body. This extensive network supplies the cells, tissues, and organs with oxygen and nutrients, while removing carbon dioxide and waste, the byproducts of respiration. The circulatory system can be thought of as a highway system that runs throughout the body.

Circulatory system is analogous to a highway system: Just as highway systems transport people and goods through a complex network, the circulatory system transports nutrients, gases, and wastes throughout the animal body.

At the core of the human circulatory system is the heart. The size of a clenched fist, the human heart is protected beneath the rib cage. Made of specialized and unique cardiac muscle, it pumps blood throughout the body and to the heart itself. Heart contractions are driven by intrinsic electrical impulses that the brain and endocrine hormones help to regulate. Understanding the heart's basic anatomy and function is important to understanding the body's circulatory and respiratory systems.

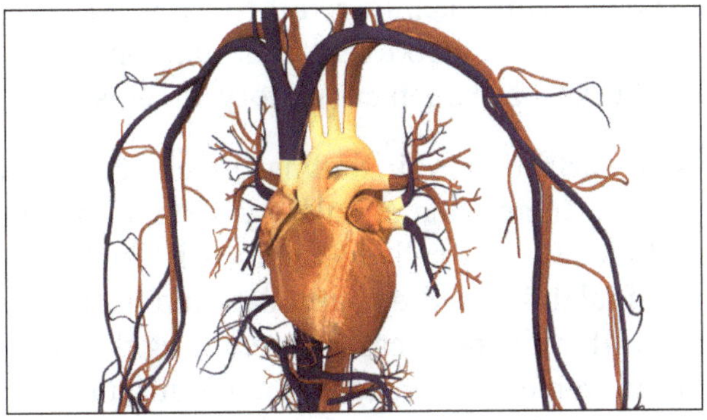

Human heart: The heart is central to the human circulatory system, as it pumps blood throughout the body.

Gas exchange is one essential function of the circulatory system. A circulatory system is not needed in organisms with no specialized respiratory organs, such as unicellular organisms, because oxygen and carbon dioxide diffuse directly between their body tissues and the external environment. However, in organisms that possess lungs and gills, oxygen must be transported from these specialized respiratory organs to the body tissues via a circulatory system. Therefore, circulatory systems have had to evolve to accommodate the great diversity of body sizes and body types present among animals.

Open and Closed Circulatory System

The circulatory system can either be open or closed, depending on whether the blood flows freely in a cavity or is contained in vessels.

Circulatory System Architecture

The circulatory system is effectively a network of cylindrical vessels (the arteries, veins, and capillaries) that emanate from a pump (the heart). In all vertebrate organisms, as well as some invertebrates, this is a closed-loop system in which the blood is not moving freely in a cavity. In a closed circulatory system, blood is contained inside blood vessels, circulating unidirectionally (in one direction) from the heart around the systemic circulatory route, then returning to the heart again.

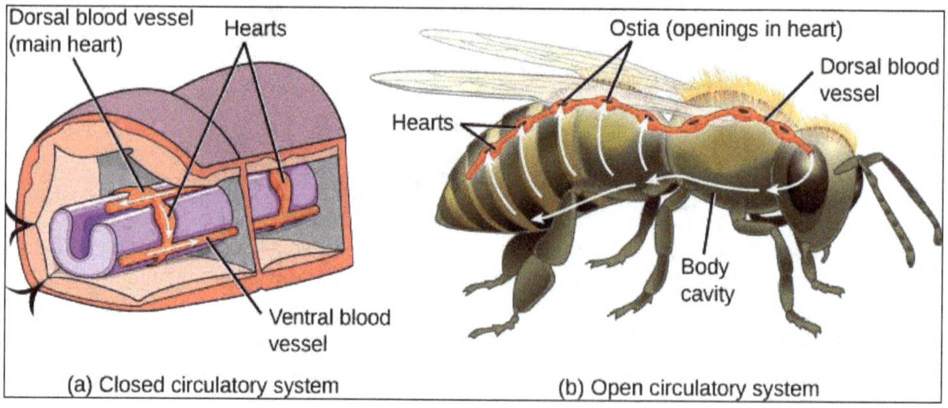

Closed and open circulatory systems: (a) In closed circulatory systems, the heart pumps blood through vessels that are separate from the interstitial fluid of the body. Most vertebrates and some invertebrates, such as this annelid earthworm, have a closed circulatory system. (b) In open circulatory systems, a fluid called hemolymph is pumped through a blood vessel that empties into the body cavity. Hemolymph returns to the blood vessel through openings called ostia. Arthropods, such as this bee and most mollusks, have open circulatory systems.

In contrast to a closed system, arthropods (including insects, crustaceans, and most mollusks) have an open circulatory system. In an open circulatory system, the blood is not enclosed in the blood vessels, but is pumped into a cavity called a hemocoel. The blood is called hemolymph because it mixes with the interstitial fluid. As the heart beats and the animal moves, the hemolymph circulates around the organs within the body cavity, reentering the heart through openings called ostia (singular: ostium). This movement allows for gas and nutrient exchange. An open circulatory system does not use as much energy to operate and maintain as a closed system; however, there is a trade-off with the amount of blood that can be moved to metabolically-active organs and tissues that require high levels of oxygen. In fact, one reason that insects with wing spans of up to two feet wide (70 cm) are not around today is probably because they were outmatched by the arrival of birds 150 million years ago. Birds, having a closed circulatory system, are thought to have moved more agilely, allowing them to obtain food faster and possibly to prey on the insects.

Digestive System

The processes in which feed materials from the environment are taken into the animal body and made available to the animal microorganism's use are complex and delightful. Nutrition in animals entails nutrient requirements by animals, the intake of feeds by animals and its utilization in the animal body.

Importance of Understanding Digestive Processes

Most feed materials are often complex and in insoluble forms as they are consumed by the animals. The aim of digestion processes therefore, is to sequentially convert the feed substances into simple and soluble materials that fit for absorption and assimilation. All animals have specialized digestive systems designed to meet their nutrient requirements. The digestive tract can be defined as a tube that extends from the mouth of the animal to its anus, constructed with dilations and constrictions throughout its dimension as to form some sections, each with its unique function in the digestive process.

Major Steps in the Digestive Process

Prehension: Snatching into or taking the feed into the mouth. Prehension is accomplished differently in different animals as follows:

- Horse: Performed by the strong and flexible upper lips and incisor teeth through jerking movements of the head or neck.
- Cattle: Uses their long, muscular tongues to pull the feed (grass) into their mouth, cutting it off between the lower incisor and the upper gum, by upward movements of the head and neck.
- Sheep: Makes the use of their mobile lips rather than tongue to gather food in their mouth.
- Horses, sheep, and cattle drink by sucking the fluid into their mouth by aid of tongue and pharynx.
- Poultry: Picks up food with their toothless beaks and passes the feed at the base of their tongue in preparation to swallowing.

Mastication or Chewing

This process follows after prehension. Mastication or chewing involves the grinding and break down of larger feed materials by animal teeth located in both the upper and lower jaws in preparation to swallowing. Differences do exists in the chewing structures of different animals in the types of teeth and their functions.

Salivary Secretion and Swallowing

After mastication or chewing the feed material in the animal mouth triggers a salivary secretions. Salivary secretion serves as a lubricant for the food to be swallowed easily. Swallowing also referred to deglutition is the pushing down of the feed from the mouth downward the esophagus. This process is accomplished by a series of muscular movements by uplifting the tongue and dilation of the pharynx. The swallowed feed (bolus) then passes into the esophagus then into the stomach. The amount of the feed swallowed varies with species.

Digestion

Comprises the breaking down of swallowed feedstuff into simple substances that can be absorbed by the animal body. Different species of animals digest certain types of feed than others. The variation in digestion is due to differences in the digestive systems in animals (monogastrics, ruminants, avian and pseudo-ruminants).

Absorption

The process or act of taking digested products into the bloodstream. Absorption of digested feed elements mostly occurs in small intestines and large intestines in some animals. Finger-like projections in the intestines called villi are the principal organs for absorption that takes up digested feed elements into the bloodstream. Absorbed elements are carried in the bloodstream to the body tissues, undergo metabolic processes and get utilized for growth, repair and energy.

Egestion

This is the final stage of digestion. Also referred to as defecation, is the process or act of discharging undigested foodstuff (Faeces) from the animal body via the anus.

Respiration

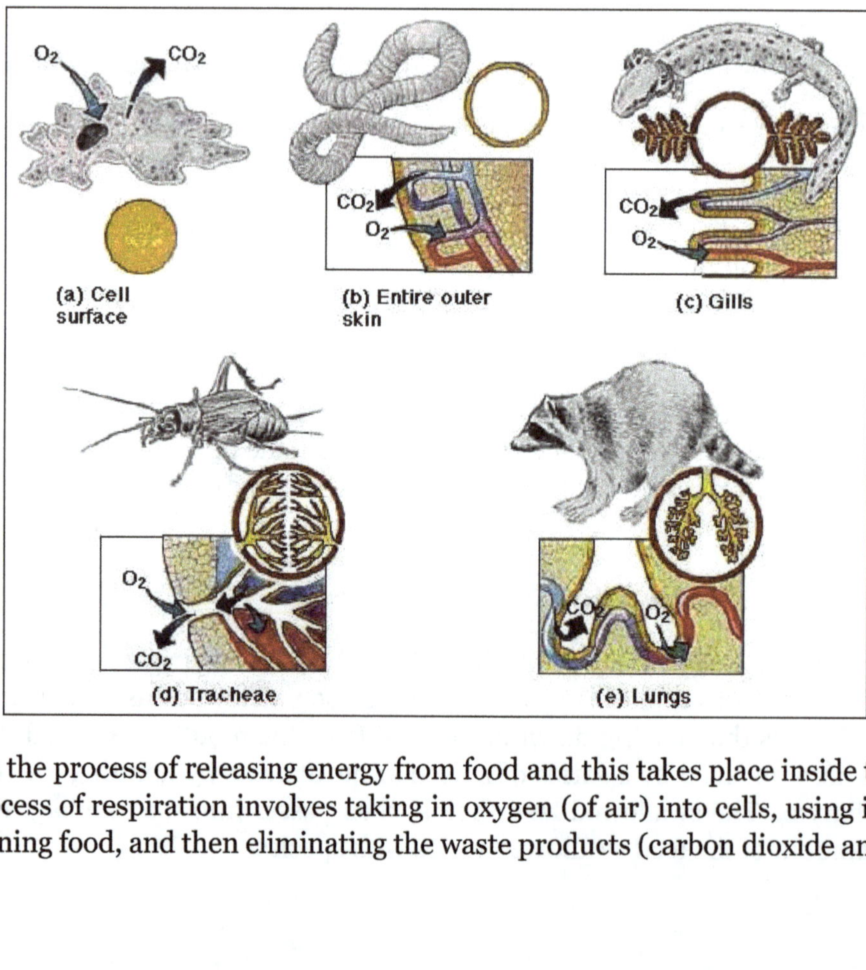

Respiration is the process of releasing energy from food and this takes place inside the cells of the body. The process of respiration involves taking in oxygen (of air) into cells, using it for releasing energy by burning food, and then eliminating the waste products (carbon dioxide and water) from the body. Respiration is an essential for life because it provides energy for carrying out all the life processes which are necessary to keep the organisms alive.

Food + Oxygen → Carbon dioxide + Water + Energy

Breathing and Respiration

The mechanism by which organisms obtain oxygen from the air and release carbon dioxide is called breathing. But on the other hand respiration is a complex process it includes breathing as well as the oxidation of food in the cells of the organism to release energy. Breathing is a physical process whereas respiration is a biochemical process of oxidation of food. Breathing process involves lungs of the organism whereas respiration involves the mitochondria in the cells where food is oxidised to release energy.

Procedure of how Energy is Released during Respiration

The energy produced during respiration is stored in the form of ATP (Adenosine TriPhosphate) molecules in the cells of the body and used by the organism as when required. To understand this we should know the meaning of ADP, ATP and Inorganic Phosphate. ADP is Adenosine Di-Phosphate; ATP is Adenosine TriPhosphate and Inorganic Phosphate is a substance which contains a phosphate group made up of phosphorous and oxygen.

The energy released during respiration is used to make ATP molecules from ADP and inorganic phosphate.

ADP (Low energy) + Phosphate + Energy (From respiration) → ATP (High energy)

This energy is stored in the cells in the form of ATP.

When the cells need energy, then ATP can be broken down using water to release energy.

ATP → ADP + Phosphate + Energy (For use in cells)

The released energy by ATP is equivalent to 30.5 kJ/mole and is used to carry out all the endodermic reactions taking place in the cells and so, ATP is known as the currency of cells.

Types of Respiration

Aerobic Respiration

The respiration which uses oxygen. In this process, the glucose food is completely broken down in to carbon dioxide and water by oxidation and this energy gets stored in the ATP molecules.

Glucose (1molecule) → (Glycolysis, in cytoplasm) forms Pyruvate (Pyruvic acid, 2 molecules)

in presence of Oxygen in mitochondria i.e Kreb's cycle) → $6CO_2 + 6H_2O$ + 38 ATP (energy)

Thus, mitochondria are the sites of aerobic respiration in the cells.

Anaerobic Respiration

The respiration which takes place without oxygen. In this, the microorganisms like yeast break down glucose (food) in to ethanol, carbon dioxide and releases energy.

Glucose (1 molecule) → undergoes Glycolysis in cytoplasm and forms

Pyruvate (Pyruvic acid, 2 molecules) → in absence of oxygen (in yeast)

Process is known as fermentation → $2C_2H_5OH$ (ethanol) + $2CO_2$ (carbon dioxide) + 2 ATP (energy)

Therefore, whole process of anaerobic respiration takes place in the cytoplasm of cells. So, we can say that all the cells do not use oxygen to produce energy. Fermentation of sugar can also be done by using anaerobic respiration of yeast.

Human beings obtain energy by aerobic respiration but sometimes anaerobic respiration takes place in our muscles during vigorous physical exercise when oxygen gets used up faster in the muscle cells than can be supplied by the human body.

Glucose (1molecule) → (Glycolysis in cytoplasm) forms Pyruvate or

pyruvic acid (2 molecules) → (in the absence of oxygen in muscle tissue)

forms 2 Lactic acid + 2 ATP (Energy)

The accumulation of lactic Acid in the muscles causes muscle cramps.

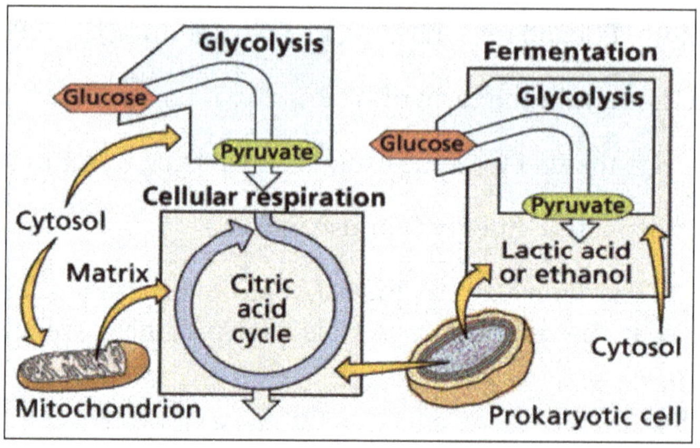

There is one similarity between aerobic and anaerobic respiration and that is energy is produced by the breakdown of food like glucose. The main differences between them are explained below:

Aerobic Respiration	Anaerobic Respiration
1. It takes place in the presence of oxygen.	1. It takes place in the absence of oxygen.
2. Complete breakdown of food occurs.	2. Partial breakdown of food occurs.
3. The end products in aerobic respiration are carbon dioxide and water.	3. The end products in anaerobic respiration may be ethanol and carbon dioxide (as in yeast plant) or lactic acid (as in animal muscles).
4. In this considerable amount of energy is produced.	4. Much less energy is produced.

Various Modes of Respiration in Animals

- In simple unicellular animals like Amoeba, respiration takes place by the simple diffusion of gases through the cell membrane. Most of the animals have, however, specific organs for respiration.
- The animals like earthworms which live in the soil use their skin to absorb oxygen from air and remove carbon dioxide. So, the respiratory organ in the earthworm is the skin.
- The aquatic animals like fish, prawns and mussels have gills as the respiratory organs which extract oxygen dissolved in water and take away carbon dioxide from the body.
- In the insects like grasshopper, cockroach, housefly and a mosquito, the tiny holes called spiracles on their body and the air tubes called tracheae are the respiratory organs.
- The respiratory organs of the land animals such as man (humans), birds, lizard, dog and frog etc. are the lungs. However, Frogs breathe both by lungs and skin.

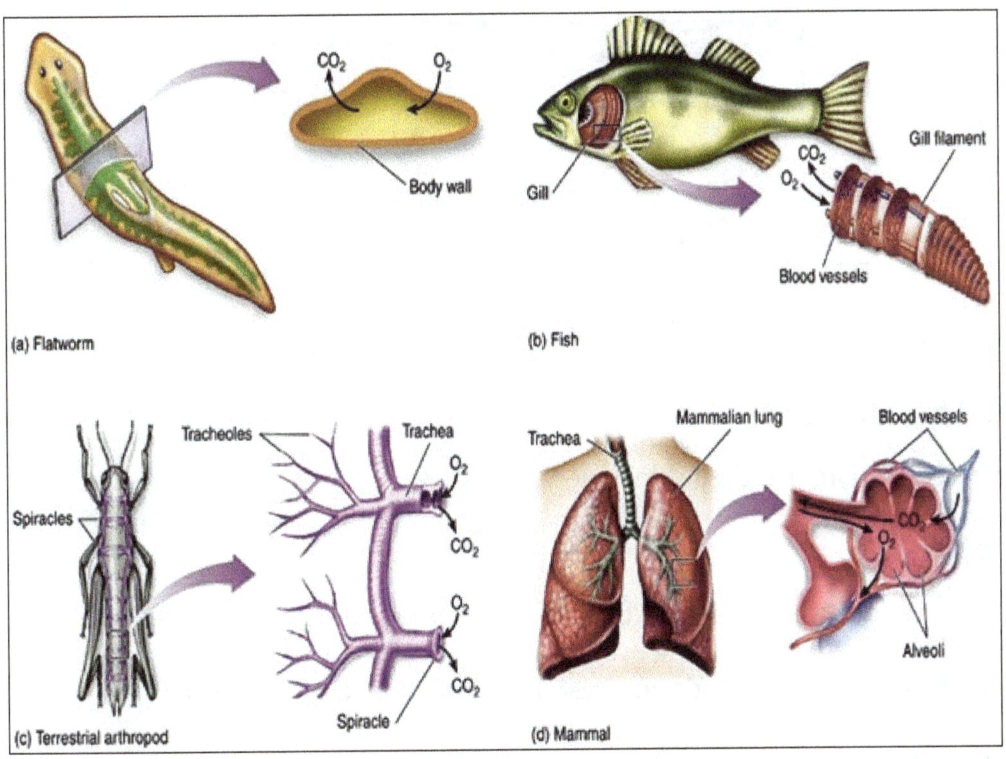

Therefore, we can say that all the respiratory organs whether skin, gills, trachea or lungs have three common features:

- All the respiratory organs have a large surface area to get enough oxygen.
- All the respiratory organs have thin walls for easy diffusion and exchange of respiratory gases.
- All the respiratory organs like skin, lungs and gills have a rich blood supply for transporting gases. But only in tracheal system of respiration, air reaches the cells directly.

Animal Anatomy

The term anatomy refers to the science that deals with the structure of animals. Animal anatomy is a broad scientific discipline which includes a wide variety of aspects, such as histological anatomy on tissues, functional anatomy to model movements, or radiological anatomy based on brain MRIs. All animals are made up of cells, some of which are specialized to carry out different functions. Simple animals, such as sponges, are made up of only a few types of cell. In more complex animals, cells are organized into tissues, such as muscles and nerves that are necessary for movement. Tissues can form organs, such as the heart, which is used to pump blood around the circulatory system.

Animal Tissue

The animal cells are grouped together to form animal tissues. These tissues vary in their structure, function, and origin. The animal tissues are divided into epithelial, connective, muscular and nervous tissues.

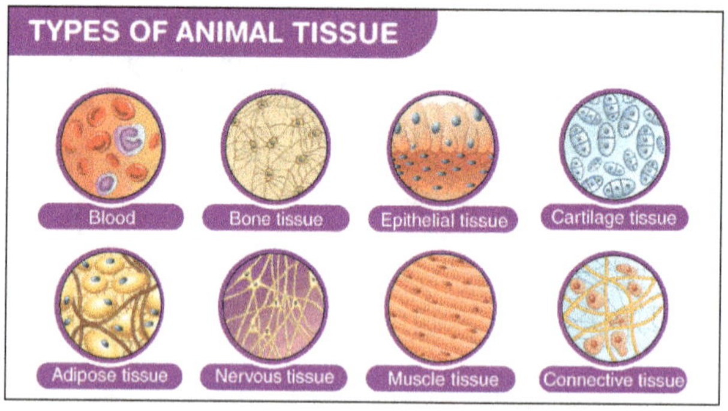

Types of Animal Tissue

The different types of animal tissues include:

Epithelial Tissue

Epithelial tissues form the protective covering and inner lining of the body and organs. These tissues were the first to evolve during evolution and were first formed during embryonic development. They develop from the ectoderm, mesoderm and endoderm of the embryo.

Characteristics of Epithelial Tissue

Following are the important characteristics of epithelial tissues:

- These can be single-layered or multi-layered.
- The tissues have the power to regenerate.
- These are held together by gap junctions, tight junctions, zonula adheren, desmosomes, or interdigitation.
- The plasma membrane of these cells is specialized into flagella, cilia, and microvilli.

Classification of Epithelial Tissues

The epithelial tissues can be classified as:

Classification	Function
Sensory epithelium	To perceive stimuli
Glandular epithelium	Secretes chemicals
Pigmented epithelium	Imparts colour in retina
Absorptive epithelium	For absorption

Connective Tissue

Connective tissues develop from the mesodermal cells of the embryo. they support and bind other tissues in the body. These are made up of three components:

- Intercellular Matrix: It is made up of mucopolysaccharide, specifically hyaluronic acid.
- Cells: The major cells include fibroblasts, adipocytes, plasma cells and mast cells.
- Fibers: Connective tissues are made up of three types of fibres, namely, collagen fibre, elastic fibre, and reticular fibre.

The connective tissues perform the following functions:

- They attach organs and tissues together.
- They store fat in the form of adipose tissues.
- They help in repairing tissues.
- They prevent the organs from mechanical shocks.
- The organs also help in defence.

Classification of Connective Tissue

The connective tissues are classified as follows:

Classification	Function
Connective Tissue Proper	Provides support and protection to the body
Vascular Tissue	Transport of materials in the body
Skeletal Tissue	It supports the body and gives it proper shape and form

Muscular Tissue

The muscular tissue develops from the mesoderm of the embryo. It is classified into three types:

- Cardiac,
- Smooth,
- Skeletal.

Muscular tissue performs the following functions:

- It helps in movement and locomotion.
- It supports the bones and other structures.
- It is responsible for peristalsis and parturition.

Classification of Muscular Tissue

The muscular tissue can be classified as:

Classification	Function
Cardiac	It helps in blood circulation and keeps the heart pumping.
Smooth	These help in peristalsis and other involuntary functions of the body.
Skeletal	Provide support, help in movement and maintain homeostasis.

Nervous Tissue

Nervous tissue makes up the peripheral and the central nervous system. It develops from the ectoderm of the embryo. It possesses the ability to initiate and transmit the nerve impulse. Its main components include:

- Neurons: These are the structural and functional unit of nervous system. It comprises an axon, cell body and dendrites.
- Neuroglia: These are special cells found in the brain and spinal cord. They provide support to the neurons and fibers.
- Neurosecretory Cells: These function as endocrine organs. They release chemical from the axons directly into blood.

Bird Anatomy

All birds have the same basic skeletal structure but with different variations in size and shape depending upon species. A bird skeleton is very unique whether the bird is flightless or adapted for flight. The skeletal structure is very lightweight and can appear fragile, however, it is actually very strong and robust. Most of the longer bones of the bird's skeleton are hollow but reinforced through a honeycombed substructure of criss-crossing for structural strength.

When birds flap their wings, it requires much strength from large muscles which need to be solidly attached to the skeleton. Flying birds have a pronounced keel or sternum, the flight muscles are attached to this. Fused bones also help the bird to withstand the stress of taking off, flying and landing. The keel is absent in flightless birds such as Penguins, Emus and Ostriches who also have solid bones instead of hollow ones.

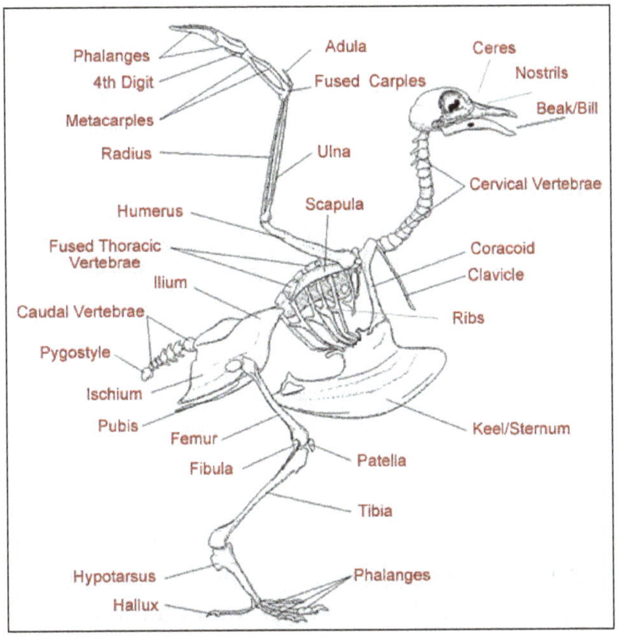

The skeletal anatomy of a bird.

The leg bones of birds are the heaviest, contributing to a low centre of gravity. This gives the bird a better balance, important in both flight and bipedal (two legged) locomotion. A bird's skeleton comprises only about 5% of its total body weight.

The Internal Anatomy of a Bird

Bird Respiratory System

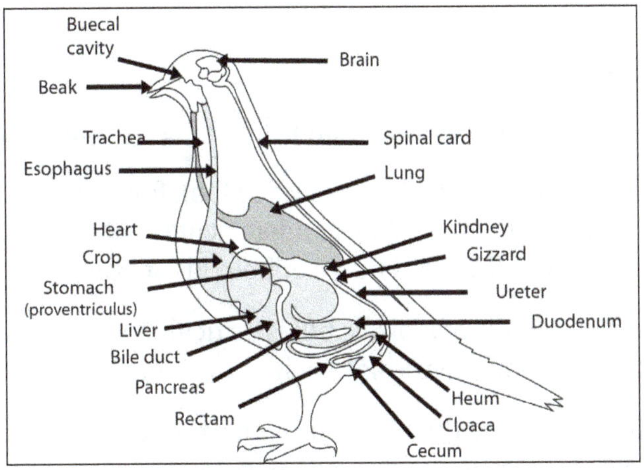

Bird Internal Anatomy.

A bird's internal anatomy, particularly its respiratory system, differs from mammals in many ways. Bird respiration is much more efficient.

- Birds do not have a diaphragm like mammals which mammals use to increase and decrease their chest cavity. Instead, birds push their sternum in and out which produces the same effect.

- Breathing rate varies depending on the size of the bird. For example, a hummingbird breathes 143 times per minute while a turkey breathes 7 times per minute. This rate increases when birds are flying. The flow of air enters through the nostrils, down the trachea and into the lungs and air sacs.
- The lungs of birds are very compact and occupy much less space than the lungs of mammals. With the help of thin-walled air sacs which extend through the body cavity and even into the bones, birds can keep a continuous flow of air through the lungs.

Digestive System

Because birds have a beak or bill and no teeth, they do not really chew their food and therefore have a specialized digestive tract.

- When a bird eats, the food firstly travels down the esophagus and into the crop which is really just an extension of the esophagus. The food is held in the crop before it can enter the proventriculus which produces acids and enzymes to help break down the food. The food then passes into the gizzard which breaks down tough foods such as seeds and nuts.
- Did you know that the Turkey has one of the most remarkable gizzards which can breakdown the toughest foods such as walnuts? It can even breakdown objects that the Turkey may have eaten such as steel needles and other metal objects.
- A bird's gizzard contains keratin to make the surface rough. Keratin is substances which makes up human hair and fingernails and is what Rhino's horns are made off.
- When the bird's food has been ground up in the gizzard, it travels through the intestines where nutrients are absorbed. Waste is then expelled through the cloaca.

Bird External Anatomy

The first step toward successful bird identification is knowing the basic parts of a bird:

- Head: The bird's head is one of the best places to look for field marks such as eye color, malar stripes, eyebrows, eye rings, eye lines and auricular patches. The crown (top) and nape (back) are also key parts of the head that can help identify a bird.
- Bill: The size, shape and color of a bird's bill is critical for identification. Also check for any curvature in the bill or unique markings such as differently colored tips or bands.
- Chin: The chin, directly below the bill, is often hard to see on many birds, but when it is a different color it can be an exceptional body part to check for identification.
- Throat: A bird's throat may be a different color from its surrounding plumage, or it may be marked with spots, streaks or lines. Malar stripes may frame the throat as well, helping set it off from the rest of a bird's body. For many birds, the chin and throat have similar colors and markings.
- Neck: The neck of a bird is hard to see on many species, since it can be relatively short and insignificant. On wading birds, however, the neck is much more prominent and can be a

good place to look for field marks. The length of the neck can also help distinguish different bird species.

- Back: A bird's back is often broad and easy to see in the right posture. Look for different colors and markings along the back that distinguish it from the neck, rump and wings.
- Chest: The chest (also called the breast) is the upright part of the bird's body between the throat and the abdomen. A bird's chest may be differently colored or marked with stripes, streaks or spots that can help with identification.
- Abdomen: The abdomen or belly of a bird extends from the bottom of the chest to the under tail coverts. The colors and markings on the abdomen may vary from the chest and flanks, making it a good feature to check for identification.
- Flanks: The flanks (sides) of a bird are located between the underside of the wings and the abdomen. In many bird species, the flanks have unique colors or markings, though depending on how the birds carry their wings, the flanks may be difficult to see.
- Wings: Birds' wings are their upper limbs used for flight. Wing bars or patches are useful field marks, as are the lengths of the wings compared to the length of the tail when the bird is perched. In flight, wing shape is also a great field mark.
- Rump: A bird's rump is the patch above the tail and low on the back. For many birds, the rump does not stand out, but some species show unique rump color patches that are useful for identification.
- Tail: The length, shape and colors of a bird's tail are important for proper identification. The tail can be held in different positions when the bird is perched or flying, however, and watching for different markings can help distinguish different birds.
- Undertail Coverts: The short feathers beneath the tail are the undertail coverts, and these feathers often show unique colors or markings that can distinguish bird species.
- Legs: Birds' legs vary in length and color, both of which can be useful field marks for proper identification. The thickness of the leg, while difficult to see on many species, can also be a clue, as can any feathering. Some raptors, for example, have heavily feathered legs that can be used to identify the birds.
- Feet: Many birds' feet are the same color as their legs, but not always. The orientation of the toes, the size of the talons and how a bird uses its feet are also useful identification characteristics.

Fish Anatomy

External Fish Anatomy

A largemouth bass shows some of the common external features that are used to describe the differences between fish that are explained in more detail below. Fish are animals that are cold-blooded, have fins and a backbone. Most fish have scales and breathe with gills. Approximately, 22,000 species of fish began evolving 480 million years ago. The largemouth bass has the typical torpedo-like (fusiform) shape associated with many fishes.

Fins are appendages used by the fish to maintain its position, move, steer and stop. They are either single fins along the centerline of the fish, such as the dorsal (back) fins, caudal (tail) fin and anal fin, or paired fins, which include the pectoral (chest) and pelvic (hip) fins. Fishes such as catfish have another fleshy lobe behind the dorsal fin, called an adipose (fat) fin. The dorsal and anal fins primarily help fish to not roll over onto their sides. The caudal fin is the main fin for propulsion to move the fish forward. The paired fins assist with steering, stopping and hovering.

Scales in most bony fishes (most freshwater fishes other than gar that have ganoid scales, and catfish which have no scales) are either ctenoid or cycloid. Ctenoid scales have jagged edges and cycloid ones have smooth rounded edges. Bass and most other fish with spines have ctenoid scales composed of connective tissue covered with calcium. Most fishes also have a very important mucus layer covering the body that helps prevent infection. Anglers should be careful not to rub this "slime" off when handling a fish that is to be released.

In many freshwater fishes the fins are supported by spines that are rigid and may be quite sharp thus playing a defensive role. Catfish have notably hard sharp fins, of which anglers should be wary. The soft dorsal and caudal fins are composed of rays, as are portions of other fins. Rays are less rigid and frequently branched.

The gills are the breathing apparatus of fish and are highly vascularized giving them their bright red cover. An operculum (gill cover) that is a flexible bony plate that protects the sensitive gills. Water is "inhaled" through the mouth, passes over the gills and is "exhaled" from beneath the operculum.

Fish can detect color. The eyes are rounder in fish than mammals because of the refractive index of water and focus is achieved by moving the lens in and out, not reshaping the lens as in mammals. Paired nostrils, or nares, in fish are used to detect odors in water and can be quite sensitive. Eels and catfish have particularly well developed senses of smell.

The mouth's shape is a good clue to what fish eat. The larger it is, the bigger the prey it can consume. Fish have a sense of taste and may sample items to taste them before swallowing if they are not obvious prey items. Most freshwater fishes in Florida are omnivorous (eating both plant and animal matter). Some are primarily piscivorous (eating mostly other fish). The imported grass carp is one of the few large fishes that are primarily herbivorous (eating plants). Fish may or may not have teeth depending on the species. Fish like chain pickerel and gar have obvious canine-shaped teeth. Other fish have less obvious teeth, such as the cardiform teeth in catfish which feel like a roughened area at the front of the mouth, or vomerine teeth that are tiny patches of teeth, for example, in the roof of a striped bass' mouth. Grass carp and other minnows have pharyngeal teeth modified from their gill arches for grinding that are located in the throat.

The lateral line is a sensory organ consisting of fluid filled sacs with hair-like sensory apparatus that are open to the water through a series of pores (creating a line along the side of the fish). The lateral line primarily senses water currents and pressure, and movement in the water. The vent is the external opening to digestive urinary and reproductive tracts. In most fish, it is immediately in front of the anal fin.

Internal Fish Anatomy

The following illustration of a largemouth bass shows some of the common internal features that are used to describe the differences between fish that are explained in more detail below.

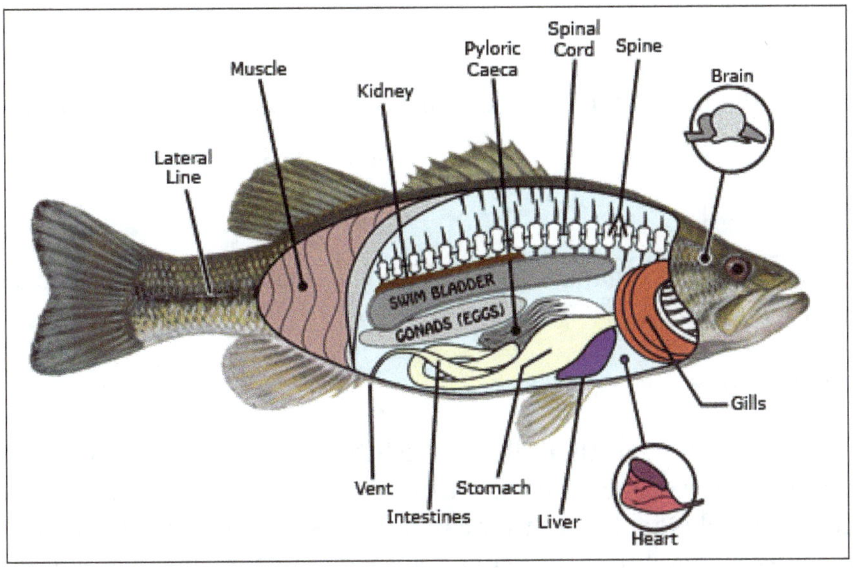

As different as a man may be from a fish, both creatures share some fascinating similarities in basic structure and function. And the closer one looks, the more complex life becomes. The smallest units of life are microscopic cells, and some organisms--such as an ameba--are no larger than a single cell. In larger multicellular creatures, individual cells that are similar in structure and perform a specific function are grouped into tissues, and tissues may be grouped into even more complex and specialized structures called organs. These organs perform the basic bodily functions such as respiration, digestion, and sensory reception. Man and fish share such organs as the brain, stomach, liver, and kidneys. Other organs appear in different forms in different organisms; for example, the lungs in humans and the gills in fish are very different but both provide the same basic function of respiration. Finally, some organs (such as the fish's swim bladder) are simply not present in man. Below are descriptions of some of the organs identified on the above diagram, along with their functions. A number of other vital organs, such as the spleen and pancreas, may also be present but are smaller and more difficult to locate. A largemouth bass destined for the frying pan makes an excellent specimen because this species is large enough for easy examination. For anglers brave enough to do some investigating while filleting their next fish, a fascinating learning experience awaits. Don't forget to examine the stomach content, as this can give clues to what the fish was feeding on, where in the water column and what lures to present. For instance, small shad probably came from open water and a shad-imitating lure may be your best bet; crayfish would suggest working a soft plastic along the bottom.

Spine

The primary structural framework upon which the fish's body is built; connects to the skull at the front of the fish and to the tail at the rear. The spine is made up of numerous vertebrae, which are hollow and house and protect the delicate spinal cord.

Spinal Cord

Connects the brain to the rest of the body and relays sensory information from the body to the brain, as well as instructions from the brain to the rest of the body.

Brain

The control center of the fish, where both automatic functions (such as respiration) and higher behaviors.

Lateral Line

One of the fish's primary sense organs; detects underwater vibrations and is capable of determining the direction of their source.

Swim (or Air) Bladder

A hollow, gas-filled balance organ that allows a fish to conserve energy by maintaining neutral buoyancy (suspending) in water. Fish caught from very deep water sometimes need to have air released from their swim bladder before they can be released and return to deep water, due to the difference in atmospheric pressure at the water's surface. (Most freshwater anglers in Florida need not concern themselves with this!) Species of fish that do not possess a swim bladder sink to the bottom if they stop swimming.

Gills

Allow a fish to breathe underwater. These are very delicate structures and should not be touched if the fish is to be released.

Kidney

Filters liquid waste materials from the blood; these wastes are then passed out of the body. The kidney is also extremely important in regulating water and salt concentrations within the fish's body, allowing certain fish species to exist in freshwater or saltwater, and in some cases (such as snook or tarpon) both.

Stomach and Intestines

Break down (digest) food and absorb nutrients. Fish such as bass that are piscivorous (eat other fish) have fairly short intestines because such food is easy to chemically break down and digest. Fish such as tilapia that are herbivorous (eat plants) require longer intestines because plant matter is usually tough and fibrous and more difficult to break down into usable components. A great deal about fish feeding habits can be determined by examining stomach contents.

Pyloric Caeca

This organ with fingerlike projections is located near the junction of the stomach and the intestines. Its function is not entirely understood, but it is known to secrete enzymes that aid in digestion, may function to absorb digested food, or do both.

Vent

The site of waste elimination from the fish's body. It is also the entry to the genital tract where eggs or sperm are released.

Liver

This important organ has a number of functions. It assists in digestion by secreting enzymes that break down fats, and also serves as a storage area for fats and carbohydrates. The liver also is important in the destruction of old blood cells and in maintaining proper blood chemistry, as well as playing a role in nitrogen (waste) excretion.

Heart

Circulates blood throughout the body. Oxygen and digested nutrients are delivered to the cells of various organs through the blood, and the blood transports waste products from the cells to the kidneys and liver for elimination.

Gonads (Reproductive Organs)

In adult female bass, the bright orange mass of eggs is unmistakable during the spawning season, but is still usually identifiable at other times of the year. The male organs, which produce milt for fertilizing the eggs, are much smaller and white but found in the same general location. The eggs (or roe) of certain fish are considered a delicacy, as in the case of caviar from sturgeon.

Muscles

Provide movement and locomotion. This is the part of the fish that is usually eaten, and composes the fillet of the fish. Horses have, on average, a skeleton of 205 bones. A significant difference in the bones contained in the horse skeleton, as compared to that of a human, is the lack of a collarbone. Their front limb system is attached to the spinal column by a powerful set of muscles, tendons and ligaments that attach the shoulder blade to the torso.

The horses legs and hooves are also unique, interesting structures. Their leg bones are proportioned differently from those of a human. For example, the body part that is called a horses 'knee' is actually the carpal bones that correspond to the human wrist. Similarly, the hock contains the bones equivalent to those in the human ankle and heel.

The lower leg bones of a horse correspond to the bones of the human hand or foot. The fetlock (incorrectly called the 'ankle') is actually the proximal sesamoid bones (bones that are typically found in locations where a tendon passes over a joint) between the cannon bones (a single equivalent to the human metacarpal (the intermediate part of the hand skeleton) or metatarsal bones (five long bones of the foot)) and the proximal phalanges (bones found in the limbs of most vertebrates), located where one finds the 'knuckles' of a human. A horse also has no muscles in its legs below the knees and hocks, only skin and hair, bone, tendons, ligaments, cartilage and the assorted specialized tissues that make up the hoof.

The skeleton gives support for the muscles, protection for the internal organs, and possesses the necessary mobility of its parts for the horse to move at various speeds or lie down or graze.

Horse Anatomy

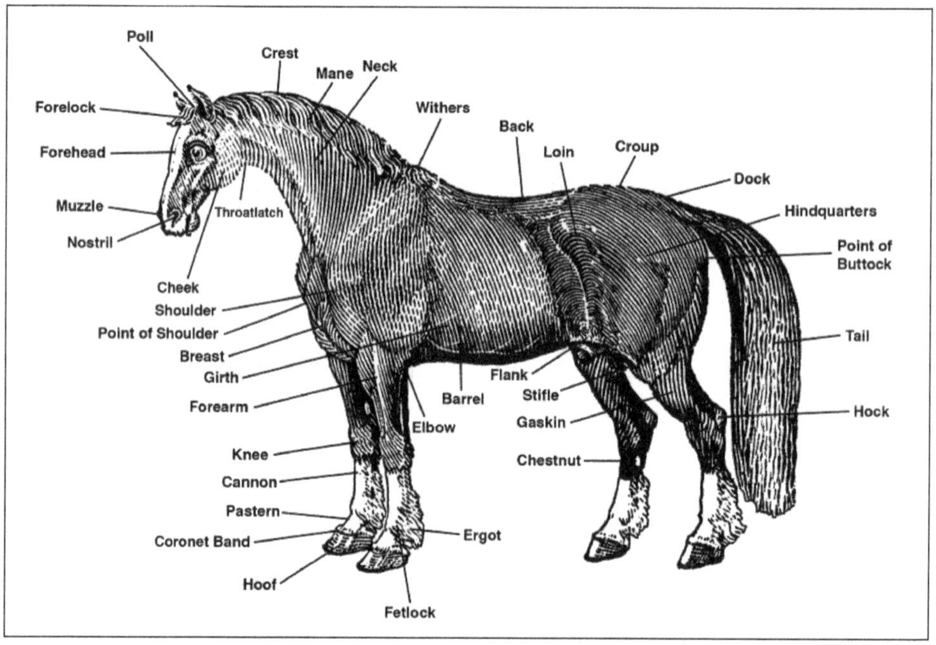

- Poll: The point where the head meets the neck, just behind the ears.
- Crest: The top portion of the neck, generally more pronounced in studs.
- Mane: Long hair which grows from the withers to the forelock and lies on either side of the neck.
- Neck: Connects at the ears and runs down to the shoulder.
- Withers: The highest point of the back, just above the shoulder blades. Horse height is measured at the withers.

- Back: Begins at the withers and extends to the last thoracic vertebrae.
- Loin: The area just behind where a saddle sits, the space from the last vertebrae to the croup.
- Croup: Begins at the top of the hip and extends down to the dock where the tail begins.
- Dock: The point where the rump and the tail connect.
- Hindquarters: The large & muscular area of the upper hind legs, behind the barrel and above the stifle.
- Point of Buttock: The rounded edge of the rump.
- Tail: Not just long, flowing hair. The tail includes the coccygeal vertebrae & the surrounding muscle and skin from which it grows.
- Hock: The bending joint (corresponds to the knee on the front legs) on the hind leg.
- Chestnut: A fleshy callous found on the inside of each leg.
- Gaskin: The large muscle just above the hock on the inside of hind legs.
- Stifle: The joint where the hind legs connect to the body.
- Flank: Area where hind legs meet the barrel just past the stifle and just before the ribcage.
- Barrel: Body of horse, essentially the area enclosed in the ribcage.
- Elbow: The joint where the front leg meets the body of the horse.
- Chestnut: A horny growth on the inside of each front leg.
- Ergot: A horny growth on the back side of the fetlock joint.
- Fetlock: The "ankle" joint (although really closer to the ball of the foot in anatomy) connecting the hoof to the leg.
- Hoof: The foot, Hard on the outside and softer on the inside. The hoof is a harder version of our fingernails.
- Coronet Band: Ring of soft tissue around the top of the hoof where it meets the skin.
- Pastern: The space between the coronet band and the fetlock.
- Cannon: The long, slender space between the knee and fetlock joints.
- Knee: The bending joint of the front legs.
- Forearm: The area between the knee and the elbow of the front legs.
- Heart Girth: Or *girth* the area just behind the elbow, where the saddle girth sits should be the largest diameter of the horse's barrel.
- Breast: The muscle mass between the forearms that cover the front of the chest.
- Point of Shoulder: The front most part of the horse's chest.

- Shoulder: The space between the withers and the point of shoulder.
- Throatlatch: Where the windpipe meets the jaw often where the strap of a bridle lays.
- Cheek: The bottom circular portion of the horses face.
- Nostril: The outside of a horse's nose.
- Muzzle: Chin, mouth and nose are all parts of the muzzle.
- Forehead: The flat space between horse's eyes.
- Forelock: A section of the mane which grows at the top of the forehead.

Chapter 6

The Biological Process of Reproduction

All living organisms procreate to maintain a continuity of their species. This biological process by which an organism replicates itself is known as reproduction and it can be either asexual or sexual. This chapter provides an extensive explanation of the various biological processes like spore formation, fertilization, fusion, fission, etc. which can be found in plant, animal and human reproduction.

Production of new individuals along a leaf margin of the air plant, Kalanchoë pinnata. The small plant in front is about 1 cm tall. The concept of "individual" is obviously stretched by this asexual reproductive process.

In biology, reproduction is the process by which new individual organisms are produced. Reproduction is a fundamental feature of all known life; each individual organism exists as the result of reproduction. Although the term reproduction encompasses a great variety of means by which organisms produce new offspring, reproductive processes can be classified into two main types: Sexual reproduction and asexual reproduction.

Sexual reproduction is a biological process by which organisms create descendants that have a combination of genetic material contributed from two different gametes (reproductive or sex cells), usually from two different organisms. The union (fusion) of these gametes produces an organism that is not genetically identical to the parent(s). Sexual reproduction generally involves contribution of genetic material to the offspring from two different organisms, but includes self-fertilization within one organism but involving fusion of two gametes. Humans are an example of a species that reproduces by sexual reproduction involving two individuals. Peanuts are a type of plant that can self-pollinate (self-fertilize).

In asexual reproduction, an individual can reproduce without involvement with another individual, there is no fusion of gametes, and the new organism produced inherits all of its chromosomes from one parent and thus is a genetically-similar or identical copy of the parent. The division of a bacterial cell into two daughter cells is an example of asexual reproduction. Asexual reproduction is not, however, limited to single-celled organisms. Most plants have the ability to reproduce asexually.

In general, more-complex organisms reproduce sexually while simpler, usually unicellular, organisms reproduce asexually. Among animals, and indeed eukaryotes in general, sexual reproduction is a nearly universal mode of reproduction. However, even lower plants, fungi, some of the protists, and bacteria likewise exhibit reproductive variances, which may be symbolized by + and - signs (rather than being called male and female), and referred to as "mating strains" or "reproductive types" or similar appellations. This polarity reflects the principle of dual characteristics, whereby existent beings exhibit the quality of positivity and negativity.

While reproductive processes can be classified into two broad groupings, species exhibit a wide spectrum of mechanisms. For example, some plants alternate between sexual and asexual reproduction. In parthenogenesis, such as found in some invertebrates (rotifers, water fleas, aphids, some bees, etc.) and some vertebrates (some amphibians, reptiles, and more rarely in birds), an embryo is produced without fertilization by a male. Generally, authors list parthenogenesis as a form of asexual reproduction because it does not involve fusion of gametes of opposite sexes, nor any exchange of genetic material from two different sources; however, some authorities (McGraw-Hill 2004) classify parthenogenesis as sexual reproduction on the basis that it involves gametes or does not produce an offspring genetically identical to the parent.

Asexual vs. Sexual Reproduction

Sexual reproduction is a near-universal mode of reproduction among eukaryotes. Furthermore, while asexual reproduction is widespread among fungi, certain protists and vascular plants, various invertebrates, and even some reptiles and amphibians, sexual reproduction is also seen in these same groups, and some state even in bacteria, which will exchange genetic material between donors (+ mating type) and recipients (- mating type). (However, many authorities, such as Mayr, consider the unidirectional lateral transfer of genetic material in bacteria to either not be reproduction, or at least not sexual reproduction, and Mayr states that sexual reproduction is unknown in prokaryotes.) Lahr contend that even in amoeboid lineages the extent of asexual reproduction in overestimated and that the evidence "demonstrates that the majority of amoeboid lineages are, contrary to popular belief, anciently sexual, and that most asexual groups have probably arisen recently and independently." Among animals, nearly all species practice sexual reproduction. Mayr notes that "above the level of the genus there are only three higher taxa of animals that consist exclusively of uniparentally reproducing clones." (An example of these higher taxa would be rotifers of the Class Bdelloidea).

Why sexual reproduction appeared and is so prevalent is a major puzzle in modern biology. Sexual reproduction has many drawbacks, since it requires far more energy than asexual reproduction. For example, in an asexual species, each member of the population is capable of bearing young, implying that an asexual population can grow more rapidly. An additional cost of sexual reproduction is that males and females must search for each other in order to mate. Evolutionary biologist and geneticist John Maynard Smith maintains that the perceived advantage for an individual organism to pass only its own entire genome to its offspring is so great that there must an advantage by at least a factor of two to explain why nearly all animal species maintain a male sex. Mayr notes that since the 1880s evolutionists have argued over the advantage of sexual reproduction and "so far, no clear-cut winner has emerged from this controversy."

General explanations for the origin and maintenance of sexual reproduction focus on the advantages conferred due to an improvement in the quality of progeny (fitness), despite reducing the overall number of offspring (two-fold cost of sex). This enhanced fitness is explained in terms of the genetic variation that is increased through sexual reproduction. Organisms that reproduce through asexual reproduction tend to grow in number exponentially. However, because they rely on mutations for variations in their DNA, all members of the species have similar vulnerabilities. Organisms that reproduce sexually yield a smaller amount of offspring, but the large amount of variation in their genes makes them less susceptible to disease or changing environmental stresses.

For example, many organisms can reproduce sexually as well as asexually. Aphids, slime molds, sea anemones, some species of starfish (by fragmentation), and many plants are examples. It is held that when environmental factors are favorable, asexual reproduction is employed to exploit suitable conditions for survival, such as an abundant food supply, adequate shelter, favorable climate, disease, optimum pH, or a proper mix of other lifestyle requirements. Populations of these organisms increase exponentially via asexual reproductive strategies to take full advantage of the rich supply resources. When food sources have been depleted, the climate becomes hostile, or individual survival is jeopardized by some other adverse change in living conditions, it is held that these organisms switch to sexual forms of reproduction. The variations found in offspring of sexual reproduction allow some individuals to be better suited for survival and provide a mechanism for selective adaptation to occur. In addition, sexual reproduction usually results in the formation of a life stage that is able to endure the conditions that threaten the offspring of an asexual parent. Thus, seeds, spores, eggs, pupae, cysts or other "over-wintering" stages of sexual reproduction ensure the survival during unfavorable times and the organism can "wait out" adverse situations until a swing back to suitability occurs.

George C. Williams introduced the *lottery principle* in 1975 to explain this basic concept, using lottery tickets as an analogy. He argued that asexual reproduction, which produces little or no genetic variety in offspring, was like buying a large number of tickets that all have the same number, limiting the chance of "winning"—that is, surviving. Sexual reproduction, he argued, was like purchasing fewer tickets but with a greater variety of numbers and therefore a greater chance of success. The point of this analogy is that since asexual reproduction does not produce genetic variations, there is little ability to quickly adapt to a changing environment. The lottery principle is less accepted these days because of evidence that asexual reproduction is more prevalent in unstable environments, the opposite of what it predicts.

Conversely, Heng proposes that the resolution to the "paradox of sex" is that sexual reproduction actually reduces the drastic genetic diversity at the genome or chromosome level, resulting in the preservation of species identity, rather than the provision of evolutionary diversity for future environmental challenges. He maintains that while genetic recombination contributes to genetic diversity, it does so secondarily and within the framework of the chromosomally defined genome. That is, the asexual process generates more diverse genomes because of the less controlled reproduction systems, while sexual reproduction generates more stable genomes.

Reproductive Strategies

There is a wide range of reproductive strategies employed by different species. Some animals, such as Homo sapiens and Northern Gannet, do not reach sexual maturity for many years after birth and even then produce few offspring. Others reproduce quickly; but, under normal circumstances, most offspring do not survive to adulthood. For example, a rabbit (mature after 8 months) can produce 10–30 offspring per year, and a fruit fly (mature after 10–14 days) can produce up to 900 offspring per year.

These two main strategies are known as K-selection (few offspring) and r-selection (many offspring). Which strategy is favored depends on a variety of circumstances. Animals with few offspring can devote more resources to the nurturing and protection of each individual offspring, thus reducing the need for a large number of offspring. On the other hand, animals with many offspring may devote less resources to each individual offspring; for these types of animals it, is common for a large number of offspring to die soon after birth, but normally enough individuals survive to maintain the population.

Other types of reproductive strategies include polycyclic animals' (reproduce intermittently throughout their lives), Semelparous organisms (reproduce only once in their lifetime, such as annual plants, which often die shortly after reproduction), and Iteroparous organisms (produce offspring in successive cycles, such as perennial plants, and thus survive over multiple seasons).

Plant Reproduction

Reproduction in plants is either asexual or sexual. Asexual reproduction in plants involves a variety of widely disparate methods for producing new plants identical in every respect to the parent. Sexual reproduction, on the other hand, depends on a complex series of basic cellular events, involving chromosomes and their genes, that take place within an elaborate sexual apparatus evolved precisely for the development of new plants in some respects different from the two parents that played a role in their production.

Sexual Reproduction

There are several different methods and processes involved in the sexual reproduction of plants. Many of the structures associated with sexual reproduction in plants are valuable commodities for humans (think fruits, berries, and vegetables). An equal number of them are "baneful" (think seasonal allergies).

Self-Pollination and Cross-Pollination

In angiosperms, pollination is defined as the placement or transfer of pollen from the anther to the stigma of the same flower or another flower. In gymnosperms, pollination involves pollen transfer from the male cone to the female cone. Upon transfer, the pollen germinates to form the pollen tube and the sperm for fertilizing the egg. Pollination has been well studied since the time of Gregor Mendel. Mendel successfully carried out self- as well as cross-pollination in garden peas while studying how characteristics were passed on from one generation to the next. Today's crops are a result of plant breeding, which employs artificial selection to produce the present-day cultivars. A case in point is today's corn, which is a result of years of breeding that started with its ancestor, teosinte. The teosinte that the ancient Mayans originally began cultivating had tiny seeds—vastly different from today's relatively giant ears of corn. Interestingly, though these two plants appear to be entirely different, the genetic difference between them is miniscule.

Pollination takes two forms: self-pollination and cross-pollination. Self-pollination occurs when the pollen from the anther is deposited on the stigma of the same flower, or another flower on the same plant. Cross-pollination is the transfer of pollen from the anther of one flower to the stigma of another flower on a different individual of the same species. Self-pollination occurs in flowers where the stamen and carpel mature at the same time, and are positioned so that the pollen can land on the flower's stigma. This method of pollination does not require an investment from the plant to provide nectar and pollen as food for pollinators.

Living species are designed to ensure survival of their progeny; those that fail become extinct. Genetic diversity is therefore required so that in changing environmental or stress conditions, some of the progeny can survive. Self-pollination leads to the production of plants with less genetic diversity, since genetic material from the same plant is used to form gametes, and eventually, the zygote. In contrast, cross-pollination—or out-crossing—leads to greater genetic diversity because the microgametophyte and megagametophyte are derived from different plants.

Because cross-pollination allows for more genetic diversity, plants have developed many ways to avoid self-pollination. In some species, the pollen and the ovary mature at different times. These flowers make self-pollination nearly impossible. By the time pollen matures and has been shed, the stigma of this flower is mature and can only be pollinated by pollen from another flower. Some flowers have developed physical features that prevent self-pollination. The primrose is one such flower. Primroses have evolved two flower types with differences in anther and stigma length: the pin-eyed flower has anthers positioned at the pollen tube's halfway point, and the thrum-eyed flower's stigma is likewise located at the halfway point. Insects easily cross-pollinate while seeking the nectar at the bottom of the pollen tube. This phenomenon is also known as heterostyly. Many plants, such as cucumber, have male and female flowers located on different parts of the plant, thus making self-pollination difficult. In yet other species, the male and female flowers are borne on different plants (dioecious). All of these are barriers to self-pollination; therefore, the plants depend on pollinators to transfer pollen. The majority of pollinators are biotic agents such as insects (like bees, flies, and butterflies), bats, birds, and other animals. Other plant species are pollinated by abiotic agents, such as wind and water.

Methods of Pollination

Pollination by Insects

Figure: Insects, such as bees, are important agents of pollination.

Bees are perhaps the most important pollinator of many garden plants and most commercial fruit trees (figure). The most common species of bees are bumblebees and honeybees. Since bees cannot see the color red, bee-pollinated flowers usually have shades of blue, yellow, or other colors. Bees collect energy-rich pollen or nectar for their survival and energy needs. They visit flowers that are open during the day, are brightly colored, have a strong aroma or scent, and have a tubular shape, typically with the presence of a nectar guide. A nectar guide includes regions on the flower petals that are visible only to bees, and not to humans; it helps to guide bees to the center of the flower, thus making the pollination process more efficient. The pollen sticks to the bees' fuzzy hair, and when the bee visits another flower, some of the pollen is transferred to the second flower. Recently, there have been many reports about the declining population of honeybees. Many flowers will remain unpollinated and not bear seed if honeybees disappear. The impact on commercial fruit growers could be devastating.

Many flies are attracted to flowers that have a decaying smell or an odor of rotting flesh. These flowers, which produce nectar, usually have dull colors, such as brown or purple. They are found on the corpse flower or voodoo lily (Amorphophallus), dragon arum (Dracunculus), and carrion flower (Stapleia, Rafflesia). The nectar provides energy, whereas the pollen provides protein. Wasps are also important insect pollinators, and pollinate many species of figs.

Butterflies, such as the monarch, pollinate many garden flowers and wildflowers, which usually occur in clusters. These flowers are brightly colored, have a strong fragrance, are open during the day, and have nectar guides to make access to nectar easier. The pollen is picked up and carried on the butterfly's limbs. Moths, on the other hand, pollinate flowers during the late afternoon and night. The flowers pollinated by moths are pale or white and are flat, enabling the moths to land. One well-studied example of a moth-pollinated plant is the yucca plant, which is pollinated by the yucca moth. The shape of the flower and moth have adapted in such a way as to allow successful pollination. The moth deposits pollen on the sticky stigma for fertilization to occur later. The female moth also deposits eggs into the ovary. As the eggs develop into larvae, they obtain food from the flower and developing seeds. Thus, both the insect and flower benefit from each other in this symbiotic relationship. The corn earworm moth and Gaura plant have a similar relationship (figure).

Figure: A corn earworm sips nectar from a night-blooming Gaura plant.

Pollination by Bats

In the tropics and deserts, bats are often the pollinators of nocturnal flowers such as agave, guava, and morning glory. The flowers are usually large and white or pale-colored; thus, they can be distinguished from the dark surroundings at night. The flowers have a strong, fruity, or musky fragrance and produce large amounts of nectar. They are naturally large and wide-mouthed to accommodate the head of the bat. As the bats seek the nectar, their faces and heads become covered with pollen, which is then transferred to the next flower.

Pollination by Birds

Many species of small birds, such as the hummingbird (figure) and sun birds, are pollinators for plants such as orchids and other wildflowers. Flowers visited by birds are usually sturdy and are oriented in such a way as to allow the birds to stay near the flower without getting their wings entangled in the nearby flowers. The flower typically has a curved, tubular shape, which allows access for the bird's beak. Brightly colored, odorless flowers that are open during the day are pollinated by birds. As a bird seeks energy-rich nectar, pollen is deposited on the bird's head and neck and is then transferred to the next flower it visits. Botanists have been known to determine the range of extinct plants by collecting and identifying pollen from 200-year-old bird specimens from the same site.

Figure: Humming birds have adaptations that allow them to reach the nectar of certain tubular flowers.

Pollination by Wind

Most species of conifers, and many angiosperms, such as grasses, maples and oaks, are pollinated by wind. Pine cones are brown and unscented, while the flowers of wind-pollinated angiosperm species are usually green, small, may have small or no petals, and produce large amounts of pollen. Unlike the typical insect-pollinated flowers, flowers adapted to pollination by wind do not produce nectar or scent. In wind-pollinated species, the microsporangia hang out of the flower, and, as the wind blows, the lightweight pollen is carried with it (figure).

Figure: A person knocks pollen from a pine tree.

The flowers usually emerge early in the spring, before the leaves, so that the leaves do not block the movement of the wind. The pollen is deposited on the exposed feathery stigma of the flower (figure).

Figure: These male (a) and female (b) catkins are from the goat willow tree (Salix caprea).

Pollination by Water

Some weeds, such as Australian sea grass and pond weeds, are pollinated by water. The pollen floats on water, and when it comes into contact with the flower, it is deposited inside the flower.

Pollination by Deception

Orchids are highly valued flowers, with many rare varieties (figure). They grow in a range of specific habitats, mainly in the tropics of Asia, South America, and Central America. At least 25,000 species of orchids have been identified.

Flowers often attract pollinators with food rewards, in the form of nectar. However, some species of orchid are an exception to this standard: they have evolved different ways to attract the desired pollinators. They use a method known as food deception, in which bright colors and perfumes are offered, but no food. Anacamptis morio, commonly known as the green-winged orchid, bears bright purple flowers and emits a strong scent. The bumblebee, its main pollinator, is attracted to the flower because of the strong scent—which usually indicates food for a bee—and in the process, picks up the pollen to be transported to another flower.

Figure: Certain orchids use food deception or sexual deception to attract pollinators. Shown here is a bee orchid (Ophrys apifera).

Other orchids use sexual deception. Chiloglottis trapeziformis emits a compound that smells the same as the pheromone emitted by a female wasp to attract male wasps. The male wasp is attracted to the scent, lands on the orchid flower, and in the process, transfers pollen. Some orchids, like the Australian hammer orchid, use scent as well as visual trickery in yet another sexual deception strategy to attract wasps. The flower of this orchid mimics the appearance of a female wasp and emits a pheromone. The male wasp tries to mate with what appears to be a female wasp, and in the process, picks up pollen, which it then transfers to the next counterfeit mate.

Double Fertilization

After pollen is deposited on the stigma, it must germinate and grow through the style to reach the ovule. The microspores, or the pollen, contain two cells: the pollen tube cell and the generative cell. The pollen tube cell grows into a pollen tube through which the generative cell travels. The germination of the pollen tube requires water, oxygen, and certain chemical signals. As it travels through the style to reach the embryo sac, the pollen tube's growth is supported by the tissues of the style. In the meantime, if the generative cell has not already split into two cells, it now divides to form two sperm cells. The pollen tube is guided by the chemicals secreted by the synergids present in the embryo sac, and it enters the ovule sac through the micropyle. Of the two sperm cells, one sperm fertilizes the egg cell, forming a diploid zygote; the other sperm fuses with the two polar nuclei, forming a triploid cell that develops into the endosperm. Together, these two fertilization events in angiosperms are known as double fertilization (figure). After fertilization is complete, no other sperm can enter. The fertilized ovule forms the seed, whereas the tissues of the ovary become the fruit, usually enveloping the seed.

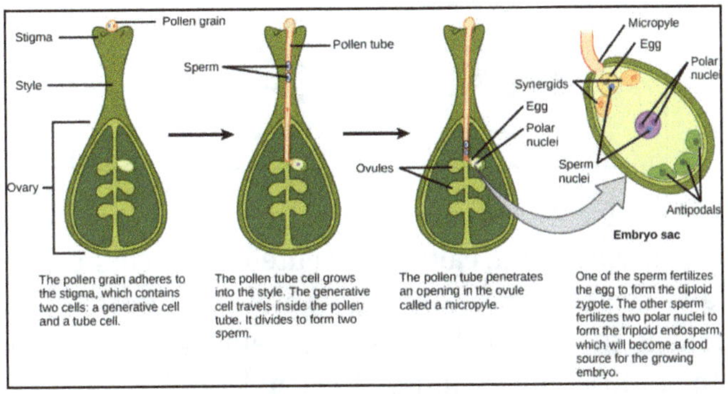

Figure: In angiosperms, one sperm fertilizes the egg to form the 2n zygote, and the other sperm fertilizes the central cell to form the 3n endosperm. This is called a double fertilization.

After fertilization, the zygote divides to form two cells: the upper cell, or terminal cell, and the lower, or basal, cell. The division of the basal cell gives rise to the suspensor, which eventually makes connection with the maternal tissue. The suspensor provides a route for nutrition to be transported from the mother plant to the growing embryo. The terminal cell also divides, giving rise to a globular-shaped proembryo (figure (a)). In dicots (eudicots), the developing embryo has a heart shape, due to the presence of the two rudimentary cotyledons (figure (b)). In non-endospermic dicots, such as Capsella bursa, the endosperm develops initially, but is then digested, and the food reserves are moved into the two cotyledons. As the embryo and cotyledons enlarge, they run out of room inside the developing seed, and are forced to bend (figure (c)). Ultimately, the embryo and cotyledons fill the seed (figure (d)), and the seed is ready for dispersal. Embryonic development is suspended after some time, and growth is resumed only when the seed germinates. The developing seedling will rely on the food reserves stored in the cotyledons until the first set of leaves begin photosynthesis.

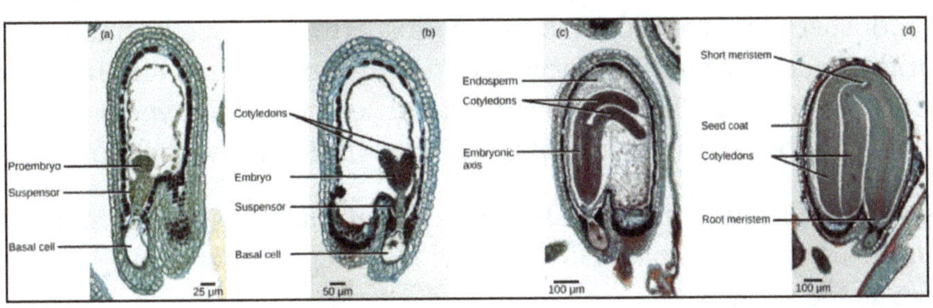

The stages of embryo development in the ovule of a shepherd's purse (Capsella bursa). After fertilization, the zygote divides to form an upper terminal cell and a lower basal cell. (a) In the first stage of development, the terminal cell divides, forming a globular pro-embryo. The basal cell also divides, giving rise to the suspensor. (b) In the second stage, the developing embryo has a heart shape due to the presence of cotyledons. (c) In the third stage, the growing embryo runs out of room and starts to bend. (d) Eventually, it completely fills the seed.

Development of a Seed

The mature ovule develops into the seed. A typical seed contains a seed coat, cotyledons, endosperm, and a single embryo (figure).

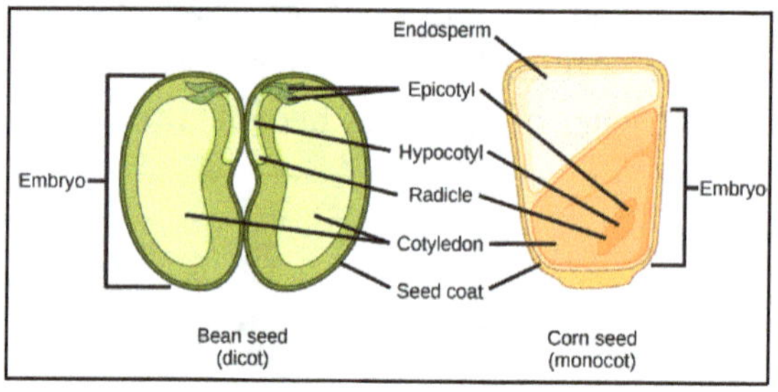

The structures of dicot and monocot seeds are shown. Dicots (left) have two cotyledons. Monocots, such as corn (right), have one cotyledon, called the scutellum; it channels nutrition to the growing embryo. Both monocot and dicot embryos have a plumule that forms the leaves, a hypocotyl that forms the stem, and a radicle that forms the root. The embryonic axis comprises everything between the plumule and the radicle, not including the cotyledon(s).

The storage of food reserves in angiosperm seeds differs between monocots and dicots. In monocots, such as corn and wheat, the single cotyledon is called a scutellum; the scutellum is connected directly to the embryo via vascular tissue (xylem and phloem). Food reserves are stored in the large endosperm. Upon germination, enzymes are secreted by the aleurone, a single layer of cells just inside the seed coat that surrounds the endosperm and embryo. The enzymes degrade the stored carbohydrates, proteins and lipids, the products of which are absorbed by the scutellum and transported via a vasculature strand to the developing embryo. Therefore, the scutellum can be seen to be an absorptive organ, not a storage organ.

The two cotyledons in the dicot seed also have vascular connections to the embryo. In endospermic dicots, the food reserves are stored in the endosperm. During germination, the two cotyledons therefore act as absorptive organs to take up the enzymatically released food reserves, much like in monocots (monocots, by definition, also have endospermic seeds). Tobacco (Nicotiana tabaccum), tomato (Solanum lycopersicum), and pepper (Capsicum annuum) are examples of endospermic dicots. In non-endospermic dicots, the triploid endosperm develops normally following double fertilization, but the endosperm food reserves are quickly remobilized and moved into the developing cotyledon for storage. The two halves of a peanut seed (Arachis hypogaea) and the split peas (Pisum sativum) of split pea soup are individual cotyledons loaded with food reserves.

The seed, along with the ovule, is protected by a seed coat that is formed from the integuments of the ovule sac. In dicots, the seed coat is further divided into an outer coat known as the testa and inner coat known as the tegmen.

The embryonic axis consists of three parts: the plumule, the radicle, and the hypocotyl. The portion of the embryo between the cotyledon attachment point and the radicle is known as the hypocotyl (hypocotyl means "below the cotyledons"). The embryonic axis terminates in a radicle (the embryonic root), which is the region from which the root will develop. In dicots, the hypocotyls extend above ground, giving rise to the stem of the plant. In monocots, the hypocotyl does not show above ground because monocots do not exhibit stem elongation. The part of the embryonic axis that projects above the cotyledons is known as the epicotyl. The plumule is composed of the epicotyl, young leaves, and the shoot apical meristem.

Upon germination in dicot seeds, the epicotyl is shaped like a hook with the plumule pointing downwards. This shape is called the plumule hook, and it persists as long as germination proceeds in the dark. Therefore, as the epicotyl pushes through the tough and abrasive soil, the plumule is protected from damage. Upon exposure to light, the hypocotyl hook straightens out, the young foliage leaves face the sun and expand, and the epicotyl continues to elongate. During this time, the radicle is also growing and producing the primary root. As it grows downward to form the tap root, lateral roots branch off to all sides, producing the typical dicot tap root system.

In monocot seeds (figure), the testa and tegmen of the seed coat are fused. As the seed germinates, the primary root emerges, protected by the root-tip covering: the coleorhiza. Next, the primary shoot emerges, protected by the coleoptile: the covering of the shoot tip. Upon exposure to light (i.e. when the plumule has exited the soil and the protective coleoptile is no longer needed), elongation of the coleoptile ceases and the leaves expand and unfold. At the other end of the embryonic axis, the primary root soon dies, while other, adventitious roots (roots that do not arise from the usual place – i.e. the root) emerge from the base of the stem. This gives the monocot a fibrous root system.

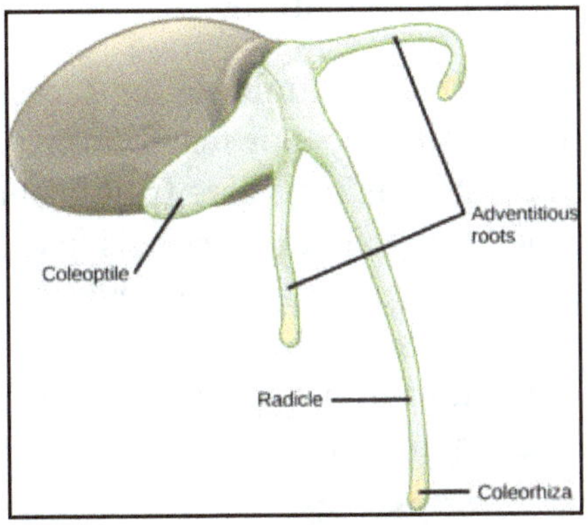

Figure: As this monocot grass seed germinates, the primary root, or radicle, emerges first, followed by the primary shoot, or coleoptile, and the adventitious roots.

Seed Germination

Many mature seeds enter a period of inactivity, or extremely low metabolic activity: a process known as dormancy, which may last for months, years or even centuries. Dormancy helps keep seeds viable during unfavorable conditions. Upon a return to favorable conditions, seed germination takes place. Favorable conditions could be as diverse as moisture, light, cold, fire, or chemical treatments. After heavy rains, many new seedlings emerge. Forest fires also lead to the emergence of new seedlings. Some seeds require vernalization (cold treatment) before they can germinate.

This guarantees that seeds produced by plants in temperate climates will not germinate until the spring. Plants growing in hot climates may have seeds that need a heat treatment in order to germinate, to avoid germination in the hot, dry summers. In many seeds, the presence of a thick seed coat retards the ability to germinate. Scarification, which includes mechanical or chemical processes to soften the seed coat, is often employed before germination. Presoaking in hot water, or passing through an acid environment, such as an animal's digestive tract, may also be employed.

Depending on seed size, the time taken for a seedling to emerge may vary. Species with large seeds have enough food reserves to germinate deep below ground, and still extend their epicotyl all the way to the soil surface. Seeds of small-seeded species usually require light as a germination cue. This ensures the seeds only germinate at or near the soil surface (where the light is greatest). If they were to germinate too far underneath the surface, the developing seedling would not have enough food reserves to reach the sunlight.

Development of Fruit and Fruit Types

After fertilization, the ovary of the flower usually develops into the fruit. Fruits are usually associated with having a sweet taste; however, not all fruits are sweet. Botanically, the term "fruit" is used for a ripened ovary. In most cases, flowers in which fertilization has taken place will develop into fruits, and flowers in which fertilization has not taken place will not. Some fruits develop from the ovary and are known as true fruits, whereas others develop from other parts of the female gametophyte and are known as accessory fruits. The fruit encloses the seeds and the developing embryo, thereby providing it with protection. Fruits are of many types, depending on their origin and texture. The sweet tissue of the blackberry, the red flesh of the tomato, the shell of the peanut, and the hull of corn (the tough, thin part that gets stuck in your teeth when you eat popcorn) are all fruits. As the fruit matures, the seeds also mature.

Fruits may be classified as simple, aggregate, multiple, or accessory, depending on their origin (figure). If the fruit develops from a single carpel or fused carpels of a single ovary, it is known as a simple fruit, as seen in nuts and beans. An aggregate fruit is one that develops from more than one carpel, but all are in the same flower: the mature carpels fuse together to form the entire fruit, as seen in the raspberry. Multiple fruit develops from an inflorescence or a cluster of flowers. An example is the pineapple, where the flowers fuse together to form the fruit. Accessory fruits (sometimes called false fruits) are not derived from the ovary, but from another part of the flower, such as the receptacle (strawberry) or the hypanthium (apples and pears).

There are four main types of fruits. Simple fruits, such as these nuts, are derived from a single ovary. Aggregate fruits, like raspberries, form from many carpels that fuse together. Multiple fruits, such as pineapple, form from a cluster of flowers called an inflorescence. Accessory fruit, like the apple, are formed from a part of the plant other than the ovary.

Fruits generally have three parts: the exocarp (the outermost skin or covering), the mesocarp (middle part of the fruit), and the endocarp (the inner part of the fruit). Together, all three are known as the pericarp. The mesocarp is usually the fleshy, edible part of the fruit; however, in some fruits, such as the almond, the endocarp is the edible part. In many fruits, two or all three of the layers are fused, and are indistinguishable at maturity. Fruits can be dry or fleshy. Furthermore, fruits can be divided into dehiscent or indehiscent types. Dehiscent fruits, such as peas, readily release their seeds, while indehiscent fruits, like peaches, rely on decay to release their seeds.

Fruit and Seed Dispersal

The fruit has a single purpose: seed dispersal. Seeds contained within fruits need to be dispersed far from the mother plant, so they may find favorable and less competitive conditions in which to germinate and grow.

Some fruit have built-in mechanisms so they can disperse by themselves, whereas others require the help of agents like wind, water, and animals (figure). Modifications in seed structure, composition, and size help in dispersal. Wind-dispersed fruit are lightweight and may have wing-like appendages that allow them to be carried by the wind. Some have a parachute-like structure to keep them afloat. Some fruits—for example, the dandelion—have hairy, weightless structures that are suited to dispersal by wind.

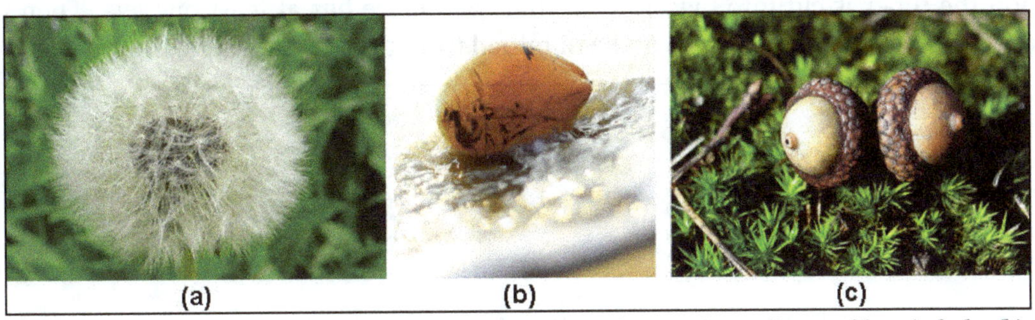

Figure: Fruits and seeds are dispersed by various means. (a) Dandelion seeds are dispersed by wind, the (b) coconut seed is dispersed by water, and the (c) acorn is dispersed by animals that cache and then forget it.

Seeds dispersed by water are contained in light and buoyant fruit, giving them the ability to float. Coconuts are well known for their ability to float on water to reach land where they can germinate. Similarly, willow and silver birches produce lightweight fruit that can float on water.

Animals and birds eat fruits, and the seeds that are not digested are excreted in their droppings some distance away. Some animals, like squirrels, bury seed-containing fruits for later use; if the squirrel does not find its stash of fruit, and if conditions are favorable, the seeds germinate. Some fruits, like the cocklebur, have hooks or sticky structures that stick to an animal's coat and are then transported to another place. Humans also play a big role in dispersing seeds when they carry fruits to new places and throw away the inedible part that contains the seeds.

Asexual Reproduction

Vegetative Propagation is an asexual method of reproduction in plants. In plants, the units of vegetative propagation such as runner, rhizome, sucker, tuber, oset, bulb are all capable of giving rise to new ospring. These structures are called vegetative propagules. These structures are called vegetative propagules. Since the formation of these structures does not involve two parents, the process involved is asexual.

Tuberous Roots: The roots of such plants have adventitious buds on their surface which sprout under favourable conditions to produce leafy shoots. In Commercial Production these sprouts are separated and planted. Thus many plants can be obtained from a single root .e.g. Sweet potato, Asparagus (Shatavari), Dahlia.

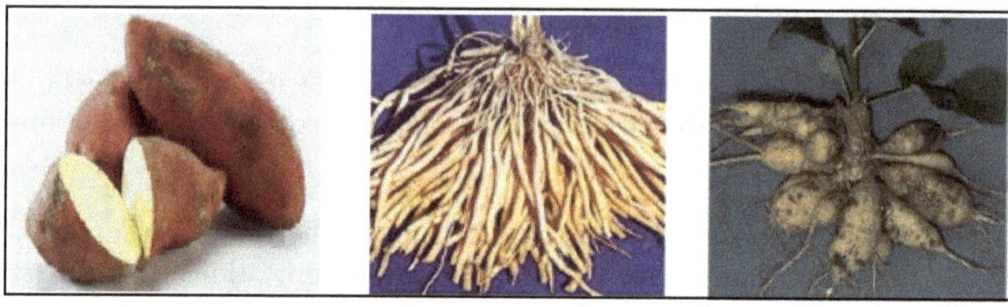

Stem Tubers: A stem tuber has many notches called 'eyes' on their surface. Each eye actually a node and consists of one or more small axillary buds and reduced scale leaves. After dormancy period the eyes which sprout under favourable conditions to produce leafy shoots. In Commercial Production the tuber is cut into pieces, such that each piece has at least one eye. Then they are grown separately. Thus many plants can be obtained from a single tuber .e.g. Potato, Ginger.

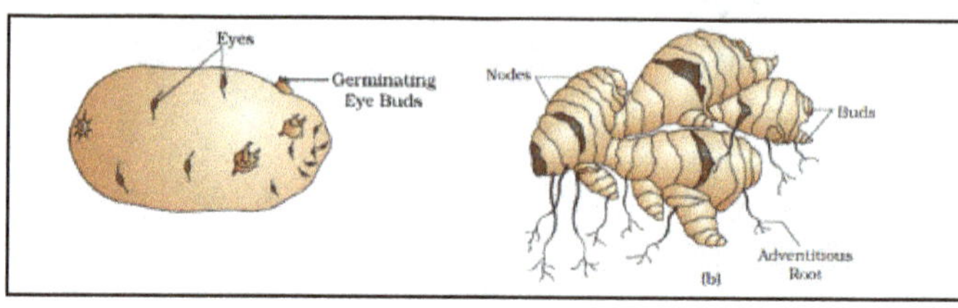

Runners: Runner develops from the lower axillary bud of stem and thin elongated cylindrical wire-like structure with long internodes. It creeps on the ground and becomes rooted at the nodes. Shoots are produced from upper sides of nodes. After getting detached from a parent, such shoots grow as an independent plant .e.g. Cynodon (doobgrass), Fragaria (strawberry), Oxalis.

Leaf: In bryophyllum, the leaf is succulent with crenate or notched margins. Adventitious buds called epiphyllous buds or foliar buds are formed at notches at the tip of lateral veins. These buds sprout and form leafy shoots and adventitious roots. When such sprout falls on wet soil, they develop into independent plants. e.g. Bryophyllum, Kalanchoe, Begonia.

Hazard due to Water Hyacinth (Terror of Bengal): One of the most invasive weeds. Grows wherever there is standing water. Drains oxygen from water-leads to the death of fishes. Introduced in India because of its pretty flowers & shape of leaves. Vegetative propagation occurs at a phenomenal rate.

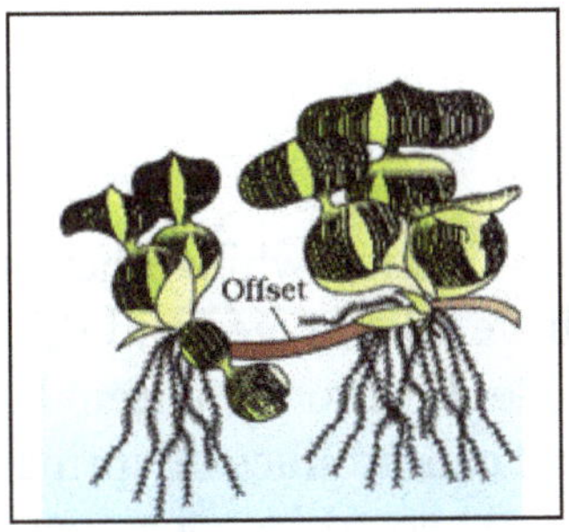

Artificial Methods of Vegetative Propagation (Asexual reproduction):

Cutting: In this method, a branch of plant part (stem, root or leaves) is cut with a node (primary meristem) on it. This piece of a branch is termed a cutting. This cutting is buried in the soil. The cutting is watered continuously. This method is the cheapest and convenient method of vegetative propagation.

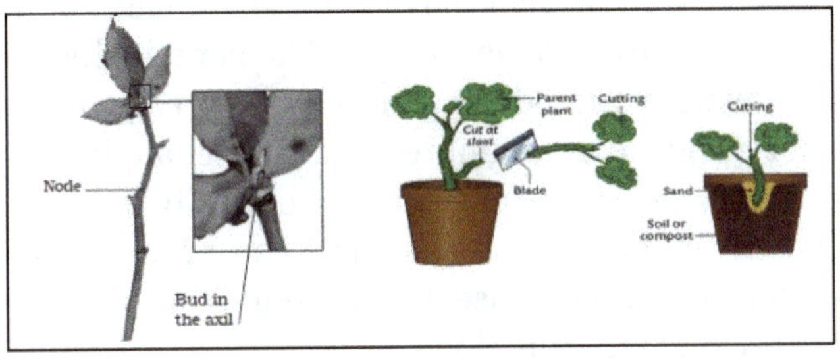

Graing: In this method, the parts of two dierent plants are joined together in such a way that they unite and continue their growth as one plant. The plant rooted to the soil and on which the part of the other plant is inserted is called a stock .The other plant which is inserted on the stock is called Scion or graft. Budding is a special case of graing in which a single bid with a small part of a bark and living tissue is grafted on another plant.

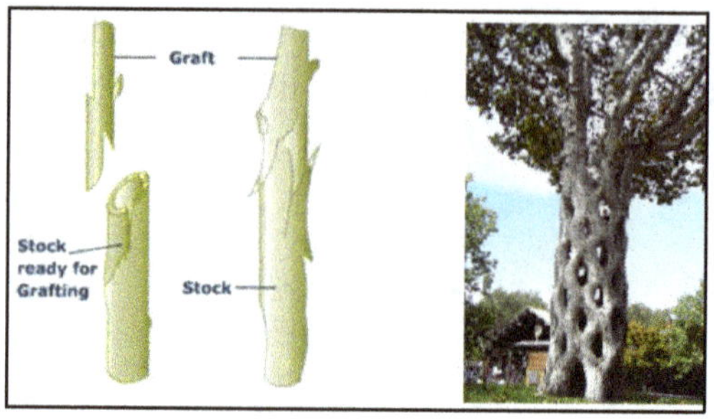

The Significance of Vegetative Propagation:

It is easy and cheaper method of multiplication . The plants like banana, pineapple and grapes which do not produce viable seeds can be propagated by vegetative propagation only. In this method, genetically similar plants are formed. The yield can be increased by graing high-yield variety on less yielding variety but which is adapted to the region. This is a rapid method of propagation particularly for the plants having a long dormancy period. It can be used to get clones of rare plants.

Animal Reproduction

Some animals produce offspring through asexual reproduction while other animals produce offspring through sexual reproduction. Both methods have advantages and disadvantages. Asexual reproduction produces offspring that are genetically identical to the parent because the offspring are all clones of the original parent. A single individual can produce offspring asexually and large numbers of offspring can be produced quickly; these are two advantages that asexually reproducing organisms have over sexually reproducing organisms. In a stable or predictable environment, asexual reproduction is an effective means of reproduction because all the offspring will be adapted to that environment. In an unstable or unpredictable environment, species that reproduce asexually may be at a disadvantage because all the offspring are genetically identical and may not be adapted to different conditions.

During sexual reproduction, the genetic material of two individuals is combined to produce genetically diverse offspring that differ from their parents. The genetic diversity of sexually produced offspring is thought to give sexually reproducing individuals greater fitness because more of their offspring may survive and reproduce in an unpredictable or changing environment. Species that reproduce sexually (and have separate sexes) must maintain two different types of individuals, males and females. Only half the population (females) can produce the offspring, so fewer offspring will be produced when compared to asexual reproduction. This is a disadvantage of sexual reproduction compared to asexual reproduction.

Asexual Reproduction

Asexual reproduction occurs in prokaryotic microorganisms (bacteria and archaea) and in many eukaryotic, single-celled and multi-celled organisms. There are several ways that animals reproduce asexually, the details of which vary among individual species.

Fission

Fission, also called binary fission, occurs in some invertebrate, multi-celled organisms. It is in some ways analogous to the process of binary fission of single-celled prokaryotic organisms. The term fission is applied to instances in which an organism appears to split itself into two parts and, if necessary, regenerate the missing parts of each new organism. For example, species of turbellarian flatworms commonly called the planarians, such as Dugesia dorotocephala, are able to separate their bodies into head and tail regions and then regenerate the missing half in each of the two new organisms. Sea anemones (Cnidaria), such as species of the genus Anthopleura, will divide along the oral-aboral axis, and sea cucumbers (Echinodermata) of the genus Holothuria, will divide into two halves across the oral-aboral axis and regenerate the other half in each of the resulting individuals.

Figure: The Anthopleura artemisia sea anemone can reproduce through fission.

Budding

Budding is a form of asexual reproduction that results from the outgrowth of a part of the body leading to a separation of the "bud" from the original organism and the formation of two individuals, one smaller than the other. Budding occurs commonly in some invertebrate animals such as hydras and corals. In hydras, a bud forms that develops into an adult and breaks away from the main body (figure).

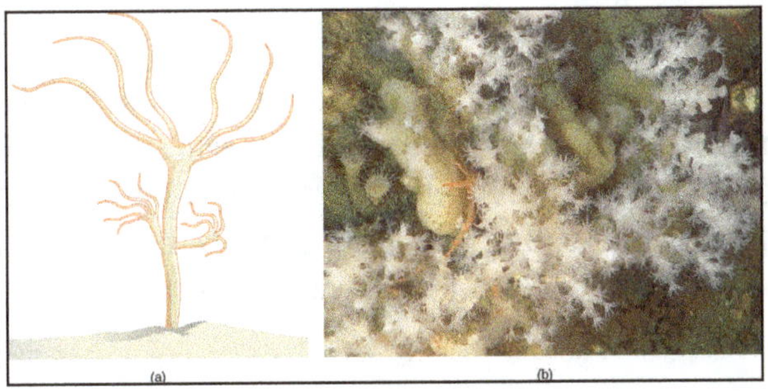

Figure (a) Hydra reproduce asexually through budding: a bud forms on the tubular body of an adult hydra, develops a mouth and tentacles, and then detaches from its parent. The new hydra is fully developed and will find its own location for attachment. (b) Some coral, such as the Lophelia pertusa shown here, can reproduce through budding. Part a: This shows a hydra, which has a stalk-like body with tentacles growing out the top. A smaller hydra is budding from the side of the stalk. Part b: This photo shows branching white coral polyps.

Fragmentation

Fragmentation is the breaking of an individual into parts followed by regeneration. If the animal is capable of fragmentation, and the parts are big enough, a separate individual will regrow from each part. Fragmentation may occur through accidental damage, damage from predators, or as a natural form of reproduction. Reproduction through fragmentation is observed in sponges, some cnidarians, turbellarians, echinoderms, and annelids. In some sea stars, a new individual can be regenerated from a broken arm and a piece of the central disc. This sea star (figure) is in the process of growing a complete sea star from an arm that has been cut off. Fisheries workers have been known to try to kill the sea stars eating their clam or oyster beds by cutting them in half and throwing them back into the ocean. Unfortunately for the workers, the two parts can each regenerate a new half, resulting in twice as many sea stars to prey upon the oysters and clams.

Figure: (a) Linckia multifora is a species of sea star that can reproduce asexually via fragmentation. In this process, (b) an arm that has been shed grows into a new sea star.

Parthenogenesis

Parthenogenesis is a form of asexual reproduction in which an egg develops into an individual without being fertilized. The resulting offspring can be either haploid or diploid, depending on the process in the species. Parthenogenesis occurs in invertebrates such as water fleas, rotifers, aphids, stick insects, and ants, wasps, and bees. Ants, bees, and wasps use parthenogenesis to produce haploid males (drones). The diploid females (workers and queens) are the result of a fertilized egg.

Some vertebrate animals—such as certain reptiles, amphibians, and fish—also reproduce through parthenogenesis. Parthenogenesis has been observed in species in which the sexes were separated in terrestrial or marine zoos. Two female Komodo dragons, a hammerhead shark, and a blacktop shark have produced parthenogenic young when the females have been isolated from males. It is possible that the asexual reproduction observed occurred in response to unusual circumstances and would normally not occur.

Sexual Reproduction

Sexual reproduction is the combination of reproductive cells from two individuals to form genetically unique offspring. The nature of the individuals that produce the two kinds of gametes can vary, having for example separate sexes or both sexes in each individual. Sex determination, the mechanism that determines which sex an individual develops into, also can vary.

Hermaphroditism

Hermaphroditism occurs in animals in which one individual has both male and female reproductive systems. Invertebrates such as earthworms, slugs, tapeworms, and snails (figure) are often hermaphroditic. Hermaphrodites may self-fertilize, but typically they will mate with another of their species, fertilizing each other and both producing offspring. Self-fertilization is more common in animals that have limited mobility or are not motile, such as barnacles and clams. Many species have specific mechanisms in place to prevent self-fertilization, because it is an extreme form of inbreeding and usually produces less fit offspring.

(a) (b)

Figure: Many (a) snails are hermaphrodites. When two individuals (b) mate, they can produce up to 100 eggs each.

Sex Determination

Mammalian sex is determined genetically by the combination of X and Y chromosomes. Individuals homozygous for X (XX) are female and heterozygous individuals (XY) are male. In mammals, the presence of a Y chromosome causes the development of male characteristics and its absence results in female characteristics. The XY system is also found in some insects and plants.

Bird sex determination is dependent on the combination of Z and W chromosomes. Homozygous for Z (ZZ) results in a male and heterozygous (ZW) results in a female. Notice that this system is the opposite of the mammalian system because in birds the female is the sex with the different sex chromosomes. The W appears to be essential in determining the sex of the individual, similar to the Y chromosome in mammals. Some fish, crustaceans, insects (such as butterflies and moths), and reptiles use the ZW system.

More complicated chromosomal sex determining systems also exist. For example, some swordtail fish have three sex chromosomes in a population.

The sex of some other species is not determined by chromosomes, but by some aspect of the environment. Sex determination in alligators, some turtles, and tuataras, for example, is dependent on the temperature during the middle third of egg development. This is referred to as environmental sex determination, or more specifically, as temperature-dependent sex determination. In many turtles, cooler temperatures during egg incubation produce males and warm temperatures produce females, while in many other species of turtles, the reverse is true. In some crocodiles and some turtles, moderate temperatures produce males and both warm and cool temperatures produce females.

Individuals of some species change their sex during their lives, switching from one to the other. If the individual is female first, it is termed protogyny or "first female," if it is male first, it is termed protandry or "first male." Oysters are born male, grow in size, and become female and lay eggs. The wrasses, a family of reef fishes, are all sequential hermaphrodites. Some of these species live in closely coordinated schools with a dominant male and a large number of smaller females. If the male dies, a female increases in size, changes sex, and becomes the new dominant male.

Fertilization

The fusion of a sperm and an egg is a process called fertilization. This can occur either inside (internal fertilization) or outside (external fertilization) the body of the female. Humans provide an example of the former, whereas frog reproduction is an example of the latter.

External Fertilization

External fertilization usually occurs in aquatic environments where both eggs and sperm are released into the water. After the sperm reaches the egg, fertilization takes place. Most external fertilization happens during the process of spawning where one or several females release their eggs and the male(s) release sperm in the same area, at the same time. The spawning may be triggered by environmental signals, such as water temperature or the length of daylight. Nearly all fish spawn, as do crustaceans (such as crabs and shrimp), mollusks (such as oysters), squid, and echinoderms (such as sea urchins and sea cucumbers). Frogs, corals, mayflies, and mosquitoes also spawn (figure).

Figure: During sexual reproduction in toads, the male grasps the female from behind and externally fertilizes the eggs as they are deposited.

Internal Fertilization

Internal fertilization occurs most often in terrestrial animals, although some aquatic animals also use this method. Internal fertilization may occur by the male directly depositing sperm in the female during mating. It may also occur by the male depositing sperm in the environment, usually in a protective structure, which a female picks up to deposit the sperm in her reproductive tract. There are three ways that offspring are produced following internal fertilization. In oviparity, fertilized eggs are laid outside the female's body and develop there, receiving nourishment from the yolk that is a part of the egg (figure(a)). This occurs in some bony fish, some reptiles, a few cartilaginous fish, some amphibians, a few mammals, and all birds. Most non-avian reptiles and insects produce leathery eggs, while birds and some turtles produce eggs with high concentrations of calcium carbonate in the shell, making them hard. Chicken eggs are an example of a hard shell. The eggs of the egg-laying mammals such as the platypus and echidna are leathery.

In ovoviparity, fertilized eggs are retained in the female, and the embryo obtains its nourishment from the egg's yolk. The eggs are retained in the female's body until they hatch inside of her, or she lays the eggs right before they hatch. This process helps protect the eggs until hatching. This occurs in some bony fish (like the platyfish Xiphophorus maculatus, figure (b)), some sharks, lizards, some snakes (garter snake Thamnophis sirtalis), some vipers, and some invertebrate animals (Madagascar hissing cockroach Gromphadorhina portentosa).

In viviparity the young are born alive. They obtain their nourishment from the female and are born in varying states of maturity. This occurs in most mammals (figure (c)), some cartilaginous fish, and a few reptiles.

(a) (b) (c)

In (a) oviparity, young develop in eggs outside the female body, as with these Harmonia axydridis beetles hatching. Some aquatic animals, like this (b) pregnant Xiphophorus maculatus are ovoviparous, with the egg developing inside the female and nutrition supplied primarily from the yolk. In mammals, nutrition is supported by the placenta, as was the case with this (c) newborn squirrel.

Human Reproduction System

Human reproductive system is an organ system by which humans reproduce and bear live offspring. Provided all organs are present, normally constructed, and functioning properly, the essential features of human reproduction are (1) liberation of an ovum, or egg, at a specific time in the reproductive cycle, (2) internal fertilization of the ovum by spermatozoa, or sperm cells, (3) transport of the fertilized ovum to the uterus, or womb, (4) implantation of the blastocyst, the early embryo developed from the fertilized ovum, in the wall of the uterus, (5) formation of a placenta and maintenance of the unborn child during the entire period of gestation, (6) birth of the child and expulsion of the placenta, and (7) suckling and care of the child, with an eventual return of the maternal organs to virtually their original state.

For this biological process to be carried out, certain organs and structures are required in both the male and the female. The source of the ova (the female germ cells) is the female ovary; that of spermatozoa (the male germ cells) is the testis. In females, the two ovaries are situated in the pelvic cavity; in males, the two testes are enveloped in a sac of skin, the scrotum, lying below and outside the abdomen. Besides producing the germ cells, or gametes, the ovaries and testes are the source of hormones that cause full development of secondary sexual characteristics and also the proper functioning of the reproductive tracts. These tracts comprise the fallopian tubes, the uterus, the vagina, and associated structures in females and the penis, the sperm channels (epididymis, ductus deferens, and ejaculatory ducts), and other related structures and glands in males. The function of the fallopian tube is to convey an ovum, which is fertilized in the tube, to the uterus, where gestation (development before birth) takes place. The function of the male ducts is to convey spermatozoa from the testis, to store them, and, when ejaculation occurs, to eject them with secretions from the male glands through the penis.

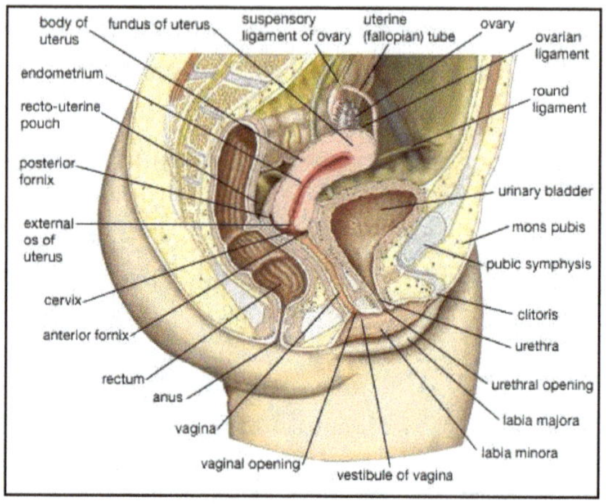

At copulation, or sexual intercourse, the erect penis is inserted into the vagina, and spermatozoa contained in the seminal fluid (semen) are ejaculated into the female genital tract. Spermatozoa then pass from the vagina through the uterus to the fallopian tube to fertilize the ovum in the outer part of the tube. Females exhibit a periodicity in the activity of their ovaries and uterus, which starts at puberty and ends at the menopause. The periodicity is manifested by menstruation at intervals of about 28 days; important changes occur in the ovaries and uterus during each reproductive, or menstrual, cycle. Periodicity, and subsequently menstruation, is suppressed during pregnancy and lactation.

Development of The Reproductive Organs

The sex of a child is determined at the time of fertilization of the ovum by the spermatozoon. The differences between a male and a female are genetically determined by the chromosomes that each possesses in the nuclei of the cells. Once the genetic sex has been determined, there normally follows a succession of changes that will result, finally, in the development of an adult male or female. There is, however, no external indication of the sex of an embryo during the first eight weeks of its life within the uterus. This is a neutral or indifferent stage during which the sex of an embryo can be ascertained only by examination of the chromosomes in its cells.

The next phase, one of differentiation, begins first in gonads that are to become testes and a week or so later in those destined to be ovaries. Embryos of the two sexes are initially alike in possessing similar duct systems linking the undifferentiated gonads with the exterior and in having similar external genitalia, represented by three simple protuberances. The embryos each have four ducts, the subsequent fate of which is of great significance in the eventual anatomical differences between men and women. Two ducts closely related to the developing urinary system are called mesonephric, or wolffian, ducts. In males each mesonephric duct becomes differentiated into four related structures: a duct of the epididymis, a ductus deferens, an ejaculatory duct, and a seminal vesicle. In females the mesonephric ducts are largely suppressed. The other two ducts, called the paramesonephric or müllerian ducts, persist, in females, to develop into the fallopian tubes, the uterus, and part of the vagina; in males they are largely suppressed. Differentiation also occurs in the primitive external genitalia, which in males become the penis and scrotum and in females the vulva (the clitoris, labia, and vestibule of the vagina).

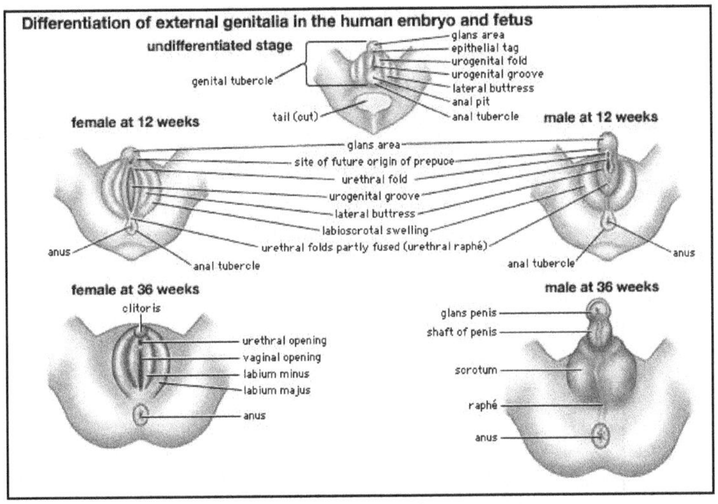

Differentiation of external genitalia in the human embryo and fetus

At birth the organs appropriate to each sex have developed and are in their adult positions but are not functioning. Various abnormalities can occur during development of sex organs in embryos, leading to hermaphroditism, pseudohermaphroditism, and other chromosomally induced conditions. During childhood until puberty there is steady growth in all reproductive organs and a gradual development of activity. Puberty marks the onset of increased activity in the sex glands and the steady development of secondary sexual characteristics.

In males at puberty the testes enlarge and become active, the external genitalia enlarge, and the capacity to ejaculate develops. Marked changes in height and weight occur as hormonal secretion from the testes increases. The larynx, or voice box, enlarges, with resultant deepening of the voice. Certain features in the skeleton, as seen in the pelvic bones and skull, become accentuated. The hair in the armpits and the pubic hair becomes abundant and thicker. Facial hair develops, as well as hair on the chest, abdomen, and limbs. Hair at the temples recedes. Skin glands become more active, especially apocrine glands (a type of sweat gland that is found in the armpits and groin and around the anus).

In females at puberty, the external genitalia enlarge and the uterus commences its periodic activity with menstruation. The breasts develop, and there is a deposition of body fat in accordance with the usual contours of the mature female. Growth of axillary (armpit) and pubic hair is more abundant, and the hair becomes thicker.

The Male Reproductive System

The male gonads are the testes; they are the source of spermatozoa and also of male sex hormones called androgens. The other genital organs are the epididymides; the ductus, or vasa, deferentia; the seminal vesicles; the ejaculatory ducts; and the penis; as well as certain accessory structures, such as the prostate and the bulbourethral (Cowper) glands. The principal functions of these structures are to transport the spermatozoa from the testes to the exterior, to allow their maturation on the way, and to provide certain secretions that help form the semen.

External Genitalia

The Penis

The penis, the male organ of copulation, is partly inside and partly outside the body. The inner part, attached to the bony margins of the pubic arch (that part of the pelvis directly in front and at the base of the trunk), is called the root of the penis. The second, or outer, portion is free, pendulous, and enveloped all over in skin; it is termed the body of the penis. The organ is composed chiefly of cavernous or erectile tissue that becomes engorged with blood to produce considerable enlargement and erection. The penis is traversed by a tube, the urethra, which serves as a passage both for urine and for semen.

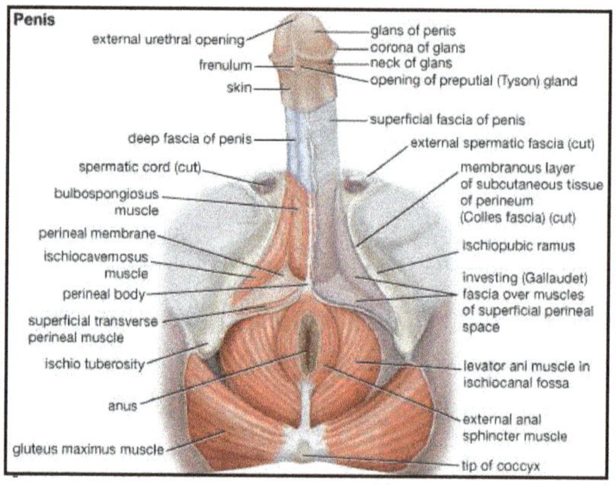

The human penis.

The body of the penis, sometimes referred to as the shaft, is cylindrical in shape when flaccid but when erect is somewhat triangular in cross section, with the angles rounded. This condition arises because the right corpus cavernosum and the left corpus cavernosum, the masses of erectile tissue, lie close together in the dorsal part of the penis, while a single body, the corpus spongiosum, which contains the urethra, lies in a midline groove on the undersurface of the corpora cavernosa. The dorsal surface of the penis is that which faces upward and backward during erection.

The slender corpus spongiosum reaches beyond the extremities of the erectile corpora cavernosa and at its outer end is enlarged considerably to form a soft, conical, sensitive structure called the glans penis. The base of the glans has a projecting margin, the corona, and the groove where the corona overhangs the corpora cavernosa is referred to as the neck of the penis. The glans is traversed by the urethra, which ends in a vertical, slitlike, external opening. The skin over the penis is thin and loosely adherent and at the neck is folded forward over the glans for a variable distance to form the prepuce or foreskin. A median fold, the frenulum of the prepuce, passes to the undersurface of the glans to reach a point just behind the urethral opening. The prepuce can usually be readily drawn back to expose the glans.

The root of the penis comprises two crura, or projections, and the bulb of the penis. The crura and the bulb are attached respectively to the edges of the pubic arch and to the perineal membrane (the fibrous membrane that forms a floor of the trunk). Each crus is an elongated structure covered by the ischiocavernosus muscle, and each extends forward, converging toward the other, to become continuous with one of the corpora cavernosa. The oval bulb of the penis lies between the two crura and is covered by the bulbospongiosus muscle. It is continuous with the corpus spongiosum. The urethra enters it on the flattened deep aspect that lies against the perineal membrane, traverses its substances, and continues into the corpus spongiosum.

The two corpora cavernosa are close to one another, separated only by a partition in the fibrous sheath that encloses them. The erectile tissue of the corpora is divided by numerous small fibrous bands into many cavernous spaces, relatively empty when the penis is flaccid but engorged with blood during erection. The structure of the tissue of the corpus spongiosum is similar to that of the corpora cavernosa, but there is more smooth muscle and elastic tissue. A deep fascia, or sheet of connective tissue, surrounding the structures in the body of the penis is prolonged to form the suspensory ligament, which anchors the penis to the pelvic bones at the midpoint of the pubic arch.

The penis has a rich blood supply from the internal pudendal artery, a branch of the internal iliac artery, which supplies blood to the pelvic structures and organs, the buttocks, and the inside of the thighs. Erection is brought about by distension of the cavernous spaces with blood, which is prevented from draining away by compression of the veins in the area.

The penis is amply supplied with sensory and autonomic (involuntary) nerves. Of the autonomic nerve fibres the sympathetic fibres cause constriction of blood vessels, and the parasympathetic fibres cause their dilation. It is usually stated that ejaculation is brought about by the sympathetic system, which at the same time inhibits the desire to urinate and also prevents the semen from entering the bladder.

The Scrotum

The scrotum is a pouch of skin lying below the pubic symphysis and just in front of the upper parts of the thighs. It contains the testes and lowest parts of the spermatic cord. A scrotal septum or partition divides the pouch into two compartments and arises from a ridge, or raphe, visible on the outside of the scrotum. The raphe turns forward onto the undersurface of the penis and is continued back onto the perineum (the area between the legs and as far back as the anus). This arrangement indicates the bilateral origin of the scrotum from two genital swellings that lie one on each side of the base of the phallus, the precursor of the penis or clitoris in the embryo. The swellings are also referred to as the labioscrotal swellings, because in females they remain separate to form the labia majora and in males they unite to form the scrotum.

The skin of the scrotum is thin, pigmented, devoid of fatty tissue, and more or less folded and wrinkled. There are some scattered hairs and sebaceous glands on its surface. Below the skin is a layer of involuntary muscle, the dartos, which can alter the appearance of the scrotum. On exposure of the scrotum to cold air or cold water, the dartos contracts and gives the scrotum a shortened, corrugated appearance; warmth causes the scrotum to become smoother, flaccid, and less closely tucked in around the testes. Beneath the dartos muscle are layers of fascia continuous with those forming the coverings of each of the two spermatic cords, which suspend the testes within the scrotum and contain each ductus deferens, the testicular blood and lymph vessels, the artery to the cremaster muscle (which draws the testes upward), the artery to each ductus deferens, the genital branch of the genitofemoral nerve, and the testicular network of nerves.

The scrotum is supplied with blood by the external pudendal branches of the femoral artery, which is the chief artery of the thigh, and by the scrotal branches of the internal pudendal artery. The veins follow the arteries. The lymphatic drainage is to the lymph nodes in the groin.

The Testes

The two testes, or testicles, which usually complete their descent into the scrotum from their point of origin on the back wall of the abdomen in the seventh month after conception, are suspended in the scrotum by the spermatic cords. Each testis is 4 to 5 cm (about 1.5 to 2 inches) long and is enclosed in a fibrous sac, the tunica albuginea. This sac is lined internally by the tunica vasculosa, containing a network of blood vessels, and is covered by the tunica vaginalis, which is a continuation of the membrane that lines the abdomen and pelvis. The tunica albuginea has extensions into each testis that act as partial partitions to divide the testis into approximately 250 compartments, or lobules.

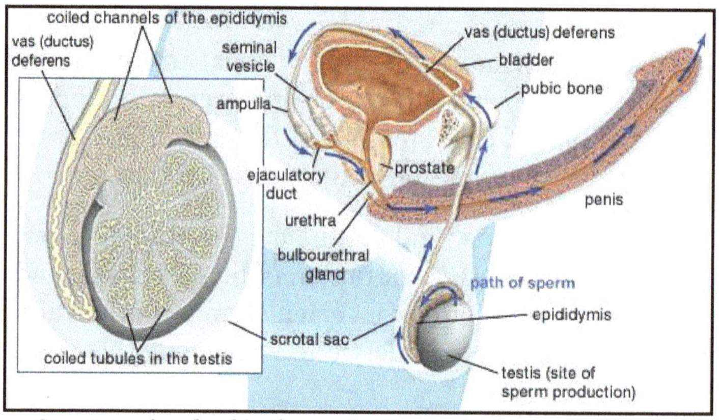

Structures involved in the production and transport of semen.

Each lobule contains one or more convoluted tubules, or narrow tubes, where sperm are formed. The tubules, if straightened, would extend about 70 cm (about 28 inches). The multistage process of sperm formation, which takes about 60 days, goes on in the lining of the tubules, starting with the spermatogonia, or primitive sperm cells, in the outermost layer of the lining. Spermatozoa (sperm) leaving the tubules are not capable of independent motion, but they undergo a further maturation process in the ducts of the male reproductive tract; the process may be continued when, after ejaculation, they pass through the female tract. Maturation of the sperm in the female tract is called capacitation.

Each spermatozoon is a slender elongated structure with a head, a neck, a middle piece, and a tail. The head contains the cell nucleus. When the spermatozoon is fully mature, it is propelled by the lashing movements of the tail.

The male sex hormone testosterone is produced by Leydig cells. These cells are located in the connective (interstitial) tissue that holds the tubules together within each lobule. The tissue becomes markedly active at puberty under the influence of the interstitial-cell-stimulating hormone of the anterior lobe of the pituitary gland; this hormone in women is called luteinizing hormone. Testosterone stimulates the male accessory sex glands (prostate, seminal vesicles) and also brings about the development of male secondary sex characteristics at puberty. The hormone may also be necessary to cause maturation of sperm and to heighten the sex drive of the male. The testis is also the source of some of the female sex hormone estrogen, which may exert an influence on pituitary activity.

Each testis is supplied with blood by the testicular arteries, which arise from the front of the aorta just below the origin of the renal (kidney) arteries. Each artery crosses the rear abdominal wall, enters the spermatic cord, passes through the inguinal canal, and enters the upper end of each testis at the back. The veins leaving the testis and epididymis form a network, which ascends into the spermatic cord. The lymph vessels, which also pass through the spermatic cord, drain to the lateral and preaortic lymph nodes. Nerve fibres to the testis accompany the vessels; they pass through the renal and aortic nerve plexuses, or networks.

Structures of the Sperm Canal

The epididymis, ductus deferens (or vas deferens), and ejaculatory ducts form the sperm canal. Together they extend from the testis to the urethra, where it lies within the prostate. Sperm are conveyed from the testis along some 20 ductules, or small ducts, which pierce the fibrous capsule to enter the head of the epididymis. The ductules are straight at first but become dilated and then much convoluted to form distinct compartments within the head of the epididymis. They each open into a single duct, the highly convoluted duct of the epididymis, which constitutes the "body" and "tail" of the structure. It is held together by connective tissue but if unraveled would be nearly 6 metres (20 feet) long. The duct enlarges and becomes thicker-walled at the lower end of the tail of the epididymis, where it becomes continuous with the ductus deferens.

The ductules from the testis have a thin muscular coat and a lining that consists of alternating groups of high columnar cells with cilia (hairlike projections) and low cells lacking cilia. The cilia assist in moving sperm toward the epididymis. In the duct of the epididymis the muscle coat is thicker and the lining is thick with tufts of large nonmotile cilia. There is some evidence that the ductules and the first portion of the duct of the epididymis remove excess fluid and extraneous debris from the testicular secretions entering these tubes. The blood supply to the epididymis is by a branch from the testicular artery given off before that vessel reaches the testis.

The ductus deferens, or vas deferens, is the continuation of the duct of the epididymis. It commences at the lower part of the tail of the epididymis and ascends along the back border of the testis to its upper pole. Then, as part of the spermatic cord, it extends to the deep inguinal ring. Separating from the other elements of the spermatic cord—the blood vessels, nerves, and lymph vessels—at the ring, the ductus deferens makes its way through the pelvis toward the base of the prostate, where it is joined by the seminal vesicle to form the ejaculatory duct. A part of the ductus that is dilated and rather tortuous, near the base of the urinary bladder, is called the ampulla.

The ductus deferens has a thick coat of smooth muscle that gives it a characteristic cordlike feel. The longitudinal muscle fibres are well developed, and peristaltic contractions (contractions in waves) move the sperm toward the ampulla. The mucous membrane lining the interior is in longitudinal folds and is mostly covered with nonciliated columnar cells, although some cells have nonmotile cilia. The ampulla is thinner-walled and probably acts as a sperm store.

Accessory Organs

The Prostate Gland, Seminal Vesicles and Bulbourethral Glands

These structures provide secretions to form the bulk of the seminal fluid of an ejaculate. The prostate gland is in the lesser or true pelvis, centred behind the lower part of the pubic arch. It lies in front of the rectum. The prostate is shaped roughly like an inverted pyramid; its base is directed upward and is immediately continuous with the neck of the urinary bladder. The urethra traverses its substance. The two ejaculatory ducts enter the prostate near the upper border of its posterior surface. The prostate is of a firm consistency, surrounded by a capsule of fibrous tissue and smooth muscle. It measures about 4 cm across, 3 cm in height, and 2 cm front to back (about 1.6 by 1.2 by 0.8 inch) and consists of glandular tissue contained in a muscular framework. It is imperfectly divided into three lobes. Two lobes at the side form the main mass and are continuous behind the urethra. In front of the urethra they are connected by an isthmus of fibromuscular tissue devoid of glands. The third, or median, lobe is smaller and variable in size and may lack glandular tissue. There are three clinically significant concentric zones of prostatic glandular tissue about the urethra. A group of short glands that are closest to the urethra and discharge mucus into its channel are subject to simple enlargement. Outside these is a ring of submucosal glands (glands from which the mucosal glands develop), and farther out is a large outer zone of long branched glands, composing the bulk of the glandular tissue. Prostate cancer is almost exclusively confined to the outer zone. The glands of the outer zone are lined by tall columnar cells that secrete prostatic fluid under the influence of androgens from the testis. The fluid is thin, milky, and slightly acidic.

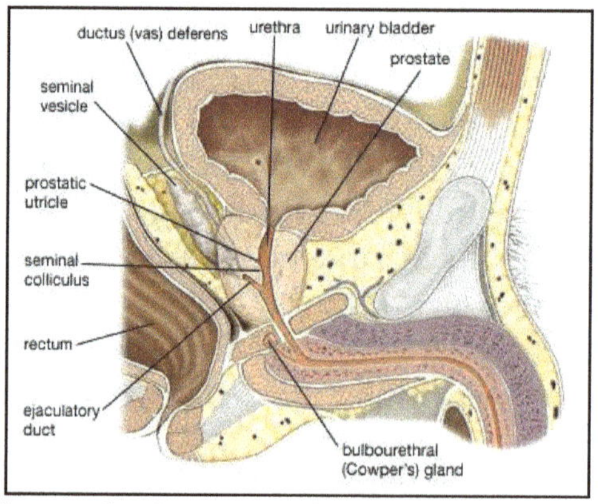

Sagittal section of the male reproductive organs, showing the prostate gland and seminal vesicles.

The seminal vesicles are two structures, about 5 cm (2 inches) in length, lying between the rectum and the base of the bladder. Their secretions form the bulk of semen. Essentially, each vesicle consists of a much-coiled tube with numerous diverticula or outpouches that extend from the main tube, the whole being held together by connective tissue. At its lower end the tube is constricted to form a straight duct or tube that joins with the corresponding ductus deferens to form the ejaculatory duct. The vesicles are close together in their lower parts, but they are separated above where they lie close to the deferent ducts. The seminal vesicles have longitudinal and circular layers of smooth muscle, and their cavities are lined with mucous membrane, which is the source of the secretions of the organs. These secretions are ejected by muscular contractions during ejaculation. The activity of the vesicles is dependent on the production of the hormone androgen by the testes. The secretion is thick, sticky, and yellowish; it contains the sugar fructose and is slightly alkaline.

The bulbourethral glands, often called Cowper glands, are pea-shaped glands that are located beneath the prostate gland at the beginning of the internal portion of the penis. The glands, which measure only about 1 cm (0.4 inch) in diameter, have slender ducts that run forward and toward the centre to open on the floor of the spongy portion of the urethra. They are composed of a network of small tubes, or tubules, and saclike structures; between the tubules are fibres of muscle and elastic tissue that give the glands muscular support. Cells within the tubules and sacs contain droplets of mucus, a thick protein compound. The fluid excreted by these glands is clear and thick and acts as a lubricant; it is also thought to function as a flushing agent that washes out the urethra before the semen is ejaculated; it may also help to make the semen less watery and to provide a suitable living environment for the sperm.

Ejaculatory Ducts

The two ejaculatory ducts lie on each side of the midline and are formed by the union of the duct of the seminal vesicle, which contributes secretions to the semen, with the end of the ductus deferens at the base of the prostate. Each duct is about 2 cm (about 0.8 inch) long and passes between a lateral and the median lobe of the prostate to reach the floor of the prostatic urethra. This part of the urethra has on its floor (or posterior wall) a longitudinal ridge called the urethral crest. On each side is a depression, the prostatic sinus, into which open the prostatic ducts. In the middle of the urethral crest is a small elevation, the colliculus seminalis, on which the opening of the prostatic utricle is found. The prostatic utricle is a short diverticulum or pouch lined by mucous membrane; it may correspond to the vagina or uterus in the female. The small openings of the ejaculatory ducts lie on each side of or just within the opening of the prostatic utricle. The ejaculatory ducts are thin-walled and lined by columnar cells.

The Female Reproductive System

The female gonads, or sexual glands, are the ovaries; they are the source of ova (eggs) and of the female sex hormones estrogens and progestogens. The fallopian, or uterine, tubes conduct ova to the uterus, which lies within the lesser or true pelvis. The uterus connects through the cervical canal with the vagina. The vagina opens into the vestibule about which lie the external genitalia, collectively known as the vulva.

External Genitalia

The female external genitalia include the structures placed about the entrance to the vagina and external to the hymen, the membrane across the entrance to the vagina. They are the mons pubis (also called the mons veneris), the labia majora and minora, the clitoris, the vestibule of the vagina, the bulb of the vestibule, and the greater vestibular glands.

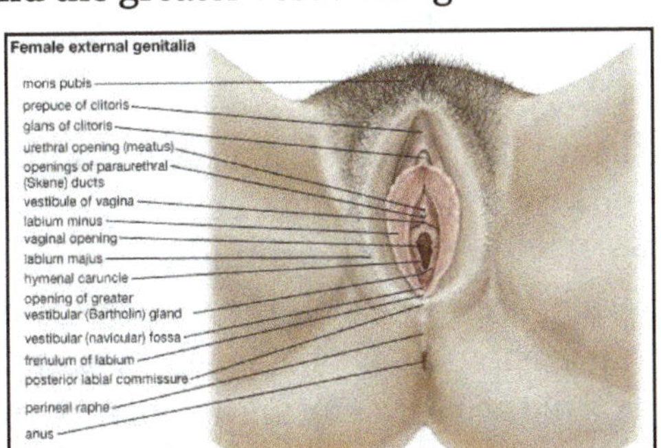

The female external genitalia.

The mons pubis is the rounded eminence, made by fatty tissue beneath the skin, lying in front of the pubic symphysis. A few fine hairs may be present in childhood; later, at puberty, they become coarser and more numerous. The upper limit of the hairy region is horizontal across the lower abdomen.

The labia majora are two marked folds of skin that extend from the mons pubis downward and backward to merge with the skin of the perineum. They form the lateral boundaries of the vulval or pudendal cleft, which receives the openings of the vagina and the urethra. The outer surface of each labium is pigmented and hairy; the inner surface is smooth but possesses sebaceous glands. The labia majora contain fat and loose connective tissue and sweat glands. They correspond to the scrotum in the male and contain tissue resembling the dartos muscle. The round ligament ends in the tissue of the labium. The labia minora are two small folds of skin, lacking fatty tissue, that extend backward on each side of the opening into the vagina. They lie inside the labia majora and are some 4 cm (about 1.5 inches) in length. In front, an upper portion of each labium minus passes over the clitoris—the structure in the female corresponding to the penis (excluding the urethra) in the male—to form a fold, the prepuce of the clitoris, and a lower portion passes beneath the clitoris to form its frenulum. The two labia minora are joined at the back across the midline by a fold that becomes stretched at childbirth. The labia minora lack hairs but possess sebaceous and sweat glands.

The clitoris is a small erectile structure composed of two corpora cavernosa separated by a partition. Partially concealed beneath the forward ends of the labia minora, it possesses a sensitive tip of spongy erectile tissue, the glans clitoridis. The external opening of the urethra is some 2.5 cm (about 1 inch) behind the clitoris and immediately in front of the vaginal opening.

The vestibule of the vagina is the cleft between the labia minora into which the urethra and vagina open. The hymen vaginae lies at the opening of the vagina: it is a thin fold of mucous membrane that varies in shape. After rupture of the hymen, the small rounded elevations that remain are known as the carunculae hymenales. The bulb of the vestibule, corresponding to the bulb of the penis, is two elongated masses of erectile tissue that lie one on each side of the vaginal opening. At their posterior ends lie the greater vestibular glands, small mucous glands that open by a duct in the groove between the hymen and each labium minus. They correspond to the bulbourethral glands of the male.

The blood supply and nerve supply of the female external genital organs are similar to those supplying corresponding structures in the male.

Internal Structures

The Vagina

The vagina (the word means "sheath") is the canal that extends from the cervix (outer end) of the uterus within the lesser pelvis down to the vestibule between the labia minora. The orifice of the vagina is guarded by the hymen. The vagina lies behind the bladder and urethra and in front of the rectum and anal canal. Its walls are collapsed; the anterior wall is some 7.5 cm (3 inches) in length, whereas the posterior wall is about 1.5 cm (0.6 inch) longer. The vagina is directed obliquely upward and backward. The axis of the vagina forms an angle of over 90° with that of the uterus. This angle varies considerably depending on conditions in the bladder, in the rectum, and during pregnancy. The cervix of the uterus projects for a short distance into the vagina and is normally pressed against its posterior wall. There are, therefore, recesses in the vagina at the back, on each side, and at the front of the cervix. These are known as the posterior fornix (behind the cervix and the largest), the lateral fornices (at the sides), and the anterior fornix (at the front of the cervix). The position of the uterus in relation to the vagina is described further in the section on the uterus.

The upper part of the posterior wall of the vagina is covered by peritoneum or membrane that is folded back onto the rectum to form the recto-uterine pouch. The lower part of the posterior vaginal wall is separated from the anal canal by a mass of tissue known as the perineal body.

The vagina has a mucous membrane and an outer smooth muscle coat closely attached to it. The mucous membrane has a longitudinal ridge in the midline of both the anterior and posterior walls. The ridges are known as the columns of the vagina; many rugae, or folds, extend from them to each side. The furrows between the rugae are more marked on the posterior wall and become especially pronounced before the birth of a child. The membrane undergoes little change during the menstrual cycle (except in its content of glycogen, a complex starchlike carbohydrate); this is in contradistinction to the situation in many mammals in which marked exfoliation (shedding of the surface cells) can occur. No glands are present in the vaginal lining, and mucus present has been secreted by the glands in the cervical canal of the uterus. The smooth muscle coat consists of an outer longitudinal layer and a less developed inner circular layer. The lower part of the vagina is surrounded by the bulbospongiosus muscle, a striped muscle attached to the perineal body.

The blood supply to the vagina is derived from several adjacent vessels, there being a vaginal artery from the internal iliac artery and also vaginal branches from the uterine, middle rectal, and internal pudendal arteries, all branches of the internal iliac artery. The nerve supply to the lower part of the vagina is from the pudendal nerve and from the inferior hypogastric and uterovaginal plexuses.

The Uterus

Uterine Structure

The uterus, or womb, is shaped like an inverted pear. It is a hollow, muscular organ with thick walls, and it has a glandular lining called the endometrium. In an adult the uterus is 7.5 cm (3 inches) long, 5 cm (2 inches) in width, and 2.5 cm (1 inch) thick, but it enlarges to four to five times this size in pregnancy. The narrower, lower end is called the cervix; this projects into the vagina. The cervix is made of fibrous connective tissue and is of a firmer consistency than the body of the uterus. The two fallopian tubes enter the uterus at opposite sides, near its top. The part of the uterus above the entrances of the tubes is called the fundus; the part below is termed the body. The body narrows toward the cervix, and a slight external constriction marks the juncture between the body and the cervix.

The uterus does not lie in line with the vagina but is usually turned forward (anteverted) to form approximately a right angle with it. The position of the uterus is affected by the amount of distension in the urinary bladder and in the rectum. Enlargement of the uterus in pregnancy causes it to rise up into the abdominal cavity, so that there is closer alignment with the vagina. The non-pregnant uterus also curves gently forward; it is said to be anteflexed. The uterus is supported and held in position by the other pelvic organs, by the muscular floor or diaphragm of the pelvis, by certain fibrous ligaments, and by folds of peritoneum. Among the supporting ligaments are two double-layered broad ligaments, each of which contains a fallopian tube along its upper free border and a round ligament, corresponding to the gubernaculum testis of the male, between its layers. Two ligaments—the cardinal (mackenrodt) ligaments—at each side of the cervix are also important in maintaining the position of the uterus.

The cavity of the uterus is remarkably small in comparison with the size of the organ. Except during pregnancy, the cavity is flattened, with front and rear walls touching, and is triangular. The triangle is inverted, with its base at the top, between the openings of the two fallopian tubes, and with its apex at the isthmus of the uterus, the opening into the cervix. The canal of the cervix is flattened from front to back and is somewhat larger in its middle part. It is traversed by two longitudinal ridges and has oblique folds stretching from each ridge in an arrangement like the branches of a tree. The cervical canal is 2.5 cm (about 1 inch) in length; its opening into the vagina is called the external os of the uterus. The external os is small, almost circular, and often depressed. After childbirth, the external os becomes bounded by lips in front and in back and is thus more slitlike. The cervical canal is lined by a mucous membrane containing numerous glands that secrete a clear, alkaline mucus. The upper part of this lining undergoes cyclical changes resembling, but not as marked as, those occurring in the body of the uterus. Numerous small cysts (nabothian cysts) are found in the cervical mucous membrane. It is from this region that cervical smears are taken in order to detect early changes indicative of cancer.

The uterus is composed of three layers of tissue. On the outside is a serous coat of peritoneum (a membrane exuding a fluid like blood minus its cells and the clotting factor fibrinogen), which partially covers the organ. In front it covers only the body of the cervix; behind it covers the body and the part of the cervix that is above the vagina and is prolonged onto the posterior vaginal wall; from there it is folded back to the rectum. At the side the peritoneal layers stretch from the margin of the uterus to each side wall of the pelvis, forming the two broad ligaments of the uterus.

The middle layer of tissue (myometrium) is muscular and comprises the greater part of the bulk of the organ. It is very firm and consists of densely packed, unstriped, smooth muscle fibres. Blood vessels, lymph vessels, and nerves are also present. The muscle is more or less arranged in three layers of fibres running in different directions. The outermost fibres are arranged longitudinally. Those of the middle layer run in all directions without any orderly arrangement; this layer is the thickest. The innermost fibres are longitudinal and circular in their arrangement.

The innermost layer of tissue in the uterus is the mucous membrane, or endometrium. It lines the uterine cavity as far as the isthmus of the uterus, where it becomes continuous with the lining of the cervical canal. The endometrium contains numerous uterine glands that open into the uterine cavity and are embedded in the cellular framework or stroma of the endometrium. Numerous blood vessels and lymphatic spaces are also present. The appearances of the endometrium vary considerably at the different stages in reproductive life. It begins to reach full development at puberty and thereafter exhibits dramatic changes during each menstrual cycle. It undergoes further changes before, during, and after pregnancy, during the menopause, and in old age. These changes are for the most part hormonally induced and controlled by the activity of the ovaries.

The Endometrium in the Menstrual Cycle

To understand the nature of the changes in the endometrium during each menstrual cycle it is usual to consider the endometrium to be composed of three layers. They blend imperceptibly but are functionally distinct: the inner two layers are shed at menstruation, and the outer or basal layer remains in position against the innermost layer of the myometrium. The three layers are called, respectively, the stratum compactum, the stratum spongiosum, and the stratum basale epidermidis. The stratum compactum is nearest to the uterine cavity and contains the lining cells and the necks of the uterine glands; its stroma is relatively dense. Superficial blood vessels lie beneath the lining cells. The stratum spongiosum is the large middle layer. It contains the main portions of uterine glands and accompanying blood vessels; the stromal cells are more loosely arranged and larger than in the stratum compactum. The stratum basale epidermidis lies against the uterine muscle; it contains blood vessels and the bases of the uterine glands. Its stroma remains relatively unaltered during the menstrual cycle.

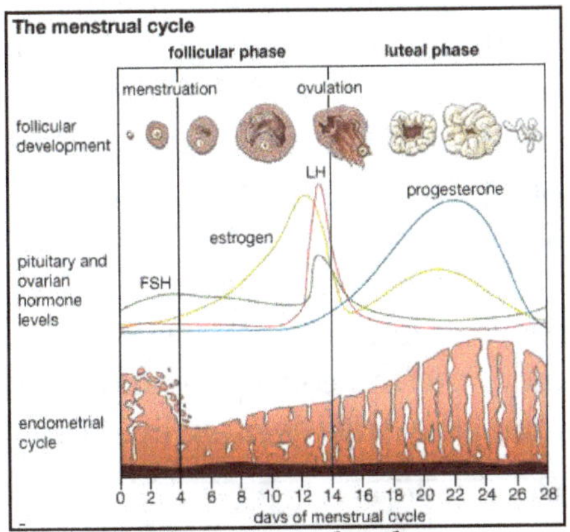

The menstrual cycle.

The menstrual cycle extends over a period of about 28 days (normal range 21–34 days), from the first day of one menstrual flow to the first day of the next. It reflects the cycle of changes occurring in the ovary, which is itself under the control of the anterior lobe of the pituitary gland. The menstrual cycle is divided into four phases: menstrual, postmenstrual, proliferative, and secretory.

The secretory phase reaches its climax about a week after ovulation. Ovulation occurs in midcycle, about 14 days before the onset of the next menstrual flow. The endometrium has been prepared and has been stimulated to a state of active secretion for the reception of a fertilized ovum. The stage has been set for the attachment of the blastocyst, derived from a fertilized ovum, to the endometrium and for its subsequent embedding. This process is called implantation; its success depends on the satisfactory preparation of the endometrium in both the proliferative and secretory phases. When implantation occurs, a hormone from certain cells of the blastocyst causes prolongation of the corpus luteum and its continued activity. This causes suppression of menstruation and results in the maintenance of the endometrium and its further stimulation by progesterone, with consequent increased thickening. The endometrium of early pregnancy is known as the decidua.

In a cycle in which fertilization of the ovum has not taken place, the secretory phase terminates in menstruation.

The endometrium needs to be in a certain state of preparedness before implantation can occur. When this stage has been passed, menstruation occurs. Repair then reestablishes an endometrium capable of being stimulated again to the critical stage when implantation can occur.

Blood Supply and Innervation

The uterus is supplied with blood by the two uterine arteries, which are branches of the internal iliac arteries, and by ovarian arteries, which connect with the ends of the uterine arteries and send branches to supply the uterus. The nerves to the uterus include the sympathetic nerve fibres, which produce contraction of uterine muscle and constriction of vessels, and parasympathetic (sacral) fibres, which inhibit muscle activity and cause dilation of blood vessels.

The Fallopian Tubes

The fallopian, or uterine, tubes carry ova from the ovaries to the cavity of the uterus. Each opens into the abdominal cavity near an ovary at one end and into the uterus at the other. Three sections of the tubes are distinguished: the funnel-shaped outer end, or infundibulum; the expanded and thin-walled intermediate portion, or ampulla; and the cordlike portion, the isthmus, that opens into the uterus. The infundibulum is fringed with irregular projections called fimbriae. One fimbria, somewhat larger than the others, is usually attached to the ovary. The opening into the abdomen is at the bottom of the infundibulum and is small. Fertilization of the ovum usually occurs in the ampulla of the tube. Normally the fertilized egg is transported to the uterus, but occasionally it may adhere to the tube and start developing as an ectopic pregnancy, or tubal pregnancy. The tube is unable to support this pregnancy, and the conceptus may be extruded through the abdominal opening or may cause rupture of the tube, with ensuing hemorrhage.

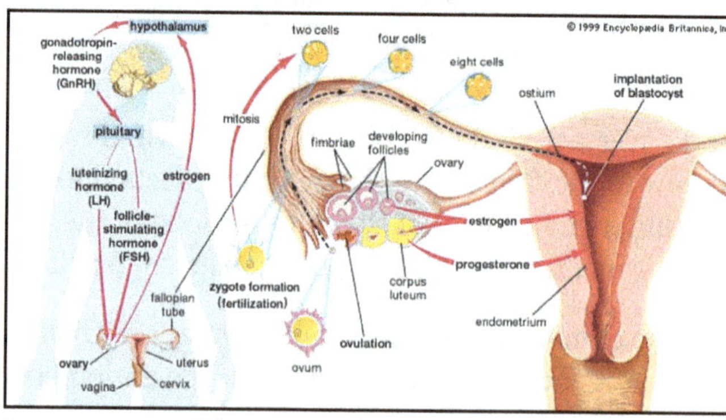

Major structures and hormones involved in the initiation of pregnancy.
At right, is the development of an egg cell (ovum) from follicle to embryo.

The fallopian tube is covered by peritoneum except on its border next to the broad ligament. There are inner circular and outer longitudinal layers of smooth muscle fibres continuous with those of the uterus. The inner lining has numerous longitudinal folds that are covered with ciliated columnar and secretory cells. Muscular contraction, movement of the hairlike cilia, and the passage of the watery secretions all probably assist in the transport of sperm to the ampulla and of a fertilized ovum toward the uterus.

The Ovaries

Ovarian Structure

The female gonads, or primary sex organs, corresponding to the testes in a male, are the two ovaries. Each is suspended by a mesentery, or fold of membrane, from the back layer of the broad ligament of the uterus. In a woman who has not been pregnant, the almond-shaped ovary lies in a vertical position against a depression, the ovarian fossa, on the side wall of the lesser pelvis. This relationship is altered during and after pregnancy. Each ovary is somewhat over 2.5 cm (1 inch) in length, 1.25 cm (0.5 inch) across, and slightly less in thickness, but the size varies much with age and with state of activity.

The mesentery of the ovary helps to keep it in position, and within this membrane lie the ovarian artery and vein, lymphatic vessels, and nerve fibres. The fallopian tube arches over the ovary and curves downward on its inner or medial surface.

Except at its hilum, the point where blood vessels and the nerve enter the ovary and where the mesentery is attached, the surface of the ovary is smooth and is covered by cubical cells. Beneath the surface, the substance of the ovary is divided into an outer portion, the cortex, and an inner portion, or medulla. The outermost part of the cortex, immediately beneath the outer covering, forms a thin connective tissue zone, the tunica albuginea. The rest of the cortex consists of stromal or framework cells, contained in a fine network of fibres, and also the follicles and corpora lutea.

The ovarian follicles, sometimes called graafian follicles, are rounded enclosures for the developing ova in the cortex near the surface of the ovary. At birth and in childhood they are present as numerous primary or undeveloped ovarian follicles. Each contains a primitive ovum, or oocyte, and each is covered by a single layer of flattened cells. As many as 700,000 primary follicles are contained in the two ovaries of a young female. Most of these degenerate before or after puberty.

Ovulation

During the onset of puberty and thereafter until menopause (except during pregnancy), there is a cyclic development of one or more follicles each month into a mature follicle. The covering layer of the primary follicle thickens and can be differentiated into an inner membrana granulosa and an outer vascularized theca interna. The cells of these layers (mostly the theca interna) produce estrogenic steroid hormones that exert their effects on the endometrium of the uterus and on other tissues. The maintenance and growth of the follicle to maturity is brought about by a follicle-stimulating hormone (FSH) from the anterior lobe of the pituitary gland. Another hormone, called luteinizing hormone (LH), from the anterior lobe, assists FSH to cause the maturing, now fluid-filled follicle to secrete estrogens. LH also causes a ripe follicle (1.0–1.5 cm [0.4–0.6 inch] in diameter) to rupture, causing the liberation of the oocyte into the peritoneal cavity and thence into the fallopian tube. This liberation of the oocyte is called ovulation; it occurs at about the midpoint of the reproductive cycle, on the 13th or 14th day of a 28-day cycle as measured from the first day of the menstrual flow.

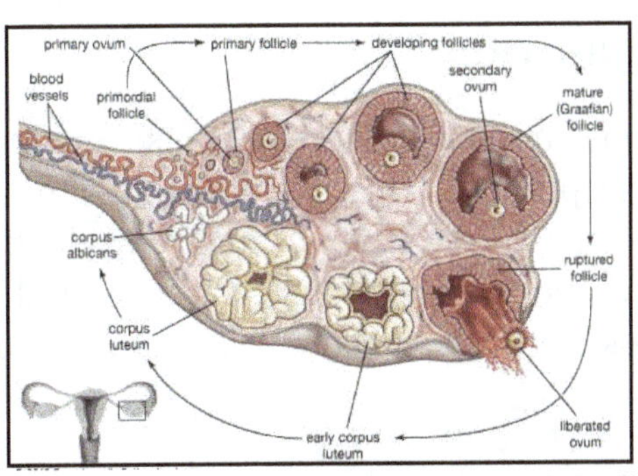

The steps of ovulation, beginning with a dormant primordial follicle that grows and matures and is eventually released from the ovary into the fallopian tube.

After ovulation the ruptured follicle collapses because of loss of its follicular fluid and rapidly becomes transformed into a soft, well-vascularized glandular structure known as the corpus luteum("yellow body"). The corpus luteum develops rapidly, becomes vascularized after about four days, and is fully established by nine days. The gland produces the steroid hormone progesterone and some estrogens. Its activity is both stimulated and maintained by luteinizing hormone. Progesterone stimulates glandular proliferation and secretion in an endometrium primed by estrogens.

While the ovarian follicle matures, the primary oocyte divides into a secondary oocyte and a small rudimentary ovum called the first polar body. This occurs at about the time when the follicle develops its cavity; the oocyte also gains a translucent acellular covering, or envelope, the zona pellucida. The secondary oocyte is liberated at ovulation; it is 120–140 micrometres in diameter and is surrounded by the zona pellucida and a few layers of cells known as the corona radiata. The final maturation of the oocyte, with the formation of the rudimentary ovum called the second polar body, occurs at the time of fertilization.

If fertilization does not occur, then the life of the corpus luteum is limited to about 14 days. Degeneration of the gland starts toward the end of this period, and menstruation occurs. The corpus luteum shrinks, fibrous tissue is formed, and it is converted into a scarlike structure called a corpus albicans, which persists for a few months.

Should fertilization occur and be followed by implantation of the blastocyst, hormones (particularly human chorionic gonadotropin) are produced by cells of the blastocyst to prolong the life of the corpus luteum. It persists in an active state for at least the first two months of pregnancy, until the placental tissue has taken over its hormone-producing function. The corpus luteum of pregnancy then also retrogresses, becoming a fibrous scar by the time of parturition.

Blood Supply and Innervation

The ovarian arteries arise from the front of the aorta in a manner similar to the testicular arteries, but at the brim of the lesser pelvis they turn down into the pelvic cavity. Passing in the suspensory ligament of the ovary, each artery reaches the broad ligament below the fallopian tube and then passes into the mesovarium to divide into branches distributed to the ovary. One branch continues in the broad ligament to anastomose with the uterine artery. The ovarian veins emerge from each ovary as a network that eventually becomes a single vein; the terminations are similar to those of the testicular veins. The nerves are derived from the ovarian nerve network on the ovarian artery.

Process of Human Reproduction

The process of reproduction in humans usually relies on sexual intercourse between a male and a female, although there are exceptions to this. Unlike many animals, humans mate throughout the year. Humans have sexual intercourse when sexual reproduction is not possible for reasons such as the use of birth control or female menopause. Practices and behaviors surrounding human reproduction vary widely across cultures, but in every case, it involves sperm, an ovum (egg), a uterus and a baby.

Gamete Production

The process of reproduction in humans begins with meiosis. In human meiosis, diploid cells with the usual 46 chromosomes divide into four haploid daughter cells, each containing 23 chromosomes. Each of these daughter cells is called a gamete. In males, this meiotic process is called spermatogenesis, and the daughter cells are sperm. In females, the process is called oogenesis, and the daughter cells are called ova. Males begin spermatogenesis at puberty and continue throughout life. Healthy young adult males produce hundreds of millions of sperm each day. This number begins to decline by their mid-20s.

Unlike males, females begin to produce gametes before they even are even born. By the fifth month in the womb, female fetuses have begun oogenesis, but the process pauses after a phase called prophase I, suspending the ova in the primary oocyte stage until puberty. 99.9 percent of a female's ova remain in the primary oocyte stage until they are eventually absorbed by the body. Millions are absorbed by the time a fetus is born, and by puberty, only 400,000 remain. For every ovulation, about 2,000 more ova are absorbed.

Sexual Intercourse

The four stages of the human sexual response cycle occur during partnered sex with people of any gender, as well as other sexual stimulation activities. The first stage is excitement, the beginning of arousal, in which blood flow increases and causes engorgement in the genitals and nipples, accompanied by an increase in heart rate, breathing rate, muscle tone and blood pressure. Next is the plateau stage, which is brief, and which involves an increase in arousal.

The third phase is orgasm, which involves waves of muscle spasms and pleasure that last several seconds. During this phase, the uterus has several contractions, and the penis has contractions at its base, causing semen, the fluid containing sperm, to ejaculate into the vagina. The last stage is resolution, during which the body relaxes to its original state.

1. Fertilization: a sperm and an egg form a zygote.

During sexual intercourse, some sperm ejaculated from the male penis swim up through the female vagina and uterus toward an oocyte (egg cell) floating in one of the uterine tubes. The sperm and the egg are gametes. They each contain half the genetic information necessary for reproduction. When a sperm cell penetrates and fertilizes an egg, that genetic information combines. The 23 chromosomes from the sperm pair with 23 chromosomes in the egg, forming a 46-chromosome cell called a zygote. The zygote starts to divide and multiply. As it travels toward the uterus it divides to become a blastocyst, which will burrow into the uterine wall.

2. The zygote becomes an embryo: development prior to and during implantation.

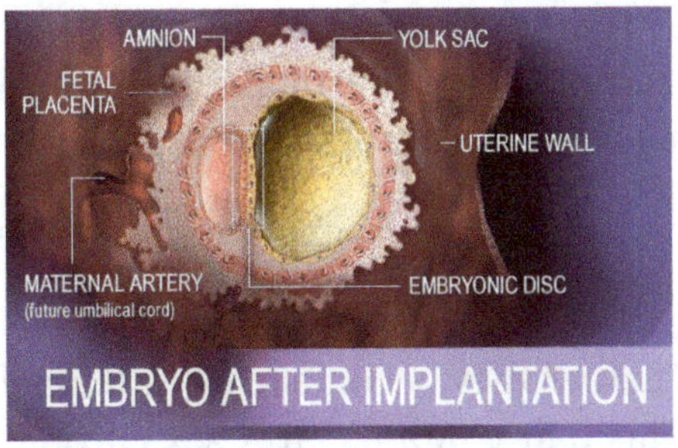

A fertilized egg, or zygote, takes about five days to reach the uterus from the uterine tube. As it moves, the zygote divides and develops into a blastocyst, with an inner mass of cells and a protective outer ring. The blastocyst attaches to the wall of the uterus and gradually implants itself into the uterine lining. During implantation, its cells differentiate further. At day 15 after conception, the cells that will form the embryo become an embryonic disc. Other cells begin to form support structures. The yolk sac, on one side of the disc, will become part of the digestive tract. On the other side, the amnion fills with fluid and will surround the embryo as it develops. Other cell groups initiate the placenta and umbilical cord, which will bring in nutrients and eliminate waste.

3. In eight weeks, the embryo develops; by the end of week 10 it becomes a fetus.

Fifteen days after conception marks the beginning of the embryonic period. The embryo contains a flat embryonic disc that now differentiates into three layers: the endoderm, the mesoderm, and the ectoderm. All organs of the human body derive from these three tissues. They begin to curve and fold and to form an oblong body. By week 4, the embryo has a distinct head and tail and a beating heart. Over the next six weeks, limbs, eyes, brain regions, and vertebrae form. Primitive versions of all body systems appear. By the end of week 10, the embryo is a fetus. (Pregnancy is often measured in terms of gestational age—age of the fetus starting with the first day of a woman's last menstrual period—and embryonic or fetal age—actual age of the growing fetus. We are referring to the gestational age of the fetus.)

4. Fetal development ends with labor, a three-stage process.

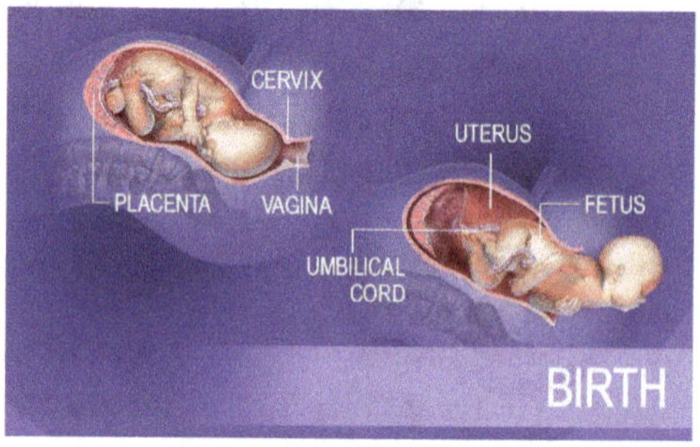

From week 10 of pregnancy, the fetus grows inside the uterus, fueled by nutrient-rich blood supplied by the umbilical cord. The placenta provides oxygen and nutrients to the fetus and removes waste products from the fetus' blood. Bones, muscles, skin, and connective tissues form. Body systems develop. Limbs and facial features take shape. Around week 36 (usually), the process of labor begins. In the first stage, dilation, hormones stimulate downward contractions of the uterine walls. The contractions push the head of the fetus against the cervix at the lower end of the uterus. The cervix dilates. In the second stage, expulsion, powerful contractions push the head and the rest of the body through the dilated cervix, and out through the vagina and the vulva. The baby is born. Further contractions expel the placenta to complete the placental stage.

5. Humans must develop male or female gonads and genitalia to be capable of reproduction.

Reproductive structures begin to form in the embryonic stage. By week 6, gonads and genitalia are present but undifferentiated. Whether they become male or female is determined by one chromosome delivered by the sperm. This pair contains an X sex chromosome from the female egg and either an X or a Y sex chromosome from the male sperm. If the chromosome pair is XY, the gonads develop into testes starting in week 7. If the chromosome pair is XX, the gonads become ovaries starting in week 8. Testes secrete testosterone, forming male genitalia around week 10. Without testosterone, female genitalia form. All reproductive structures are in place at birth or shortly after. At puberty, an increase in sex hormones will grow them to their adult size and reproductive capability.

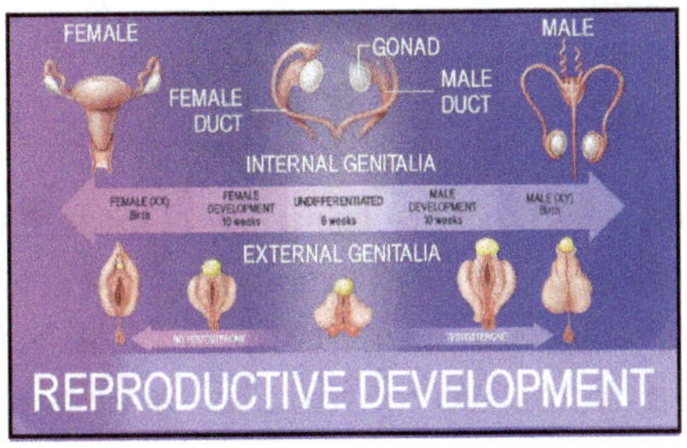

Hormonal Control of Reproduction

The human male and female reproductive cycles are controlled by the interaction of hormones from the hypothalamus and anterior pituitary with hormones from reproductive tissues and organs. In both sexes, the hypothalamus monitors and causes the release of hormones from the anterior pituitary gland. When the reproductive hormone is required, the hypothalamus sends a gonadotropin-releasing hormone (GnRH) to the anterior pituitary. This causes the release of follicle stimulating hormone (FSH) and luteinizing hormone (LH) from the anterior pituitary into the blood. Although these hormones are named after their functions in female reproduction, they are produced in both sexes and play important roles in controlling reproduction. Other hormones have specific functions in the male and female reproductive systems.

Male Hormones

At the onset of puberty, the hypothalamus causes the release of FSH and LH into the male system for the first time. FSH enters the testes and stimulates the sertoli cells located in the walls of the seminiferous tubules to begin promoting spermatogenesis (figure). LH also enters the testes and stimulates the interstitial cells of Leydig, located in between the walls of the seminiferous tubules, to make and release testosterone into the testes and the blood.

Testosterone stimulates spermatogenesis. This hormone is also responsible for the secondary sexual characteristics that develop in the male during adolescence. The secondary sex characteristics in males include a deepening of the voice, the growth of facial, axillary, and pubic hair, an increase in muscle bulk, and the beginnings of the sex drive.

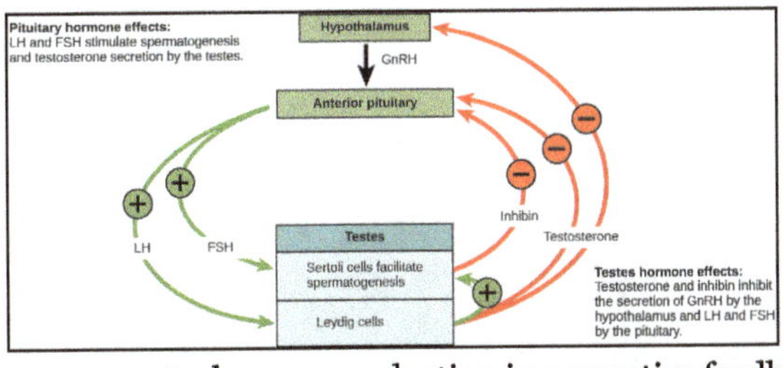

Figure: Hormones control sperm production in a negative feedback system.

A negative feedback system occurs in the male with rising levels of testosterone acting on the hypothalamus and anterior pituitary to inhibit the release of GnRH, FSH, and LH. In addition, the sertoli cells produce the hormone inhibin, which is released into the blood when the sperm count is too high. This inhibits the release of GnRH and FSH, which will cause spermatogenesis to slow down. If the sperm count reaches a low of 20 million/mL, the Sertoli cells cease the release of inhibin, and the sperm count increases.

Female Hormones

The control of reproduction in females is more complex. The female reproductive cycle is divided into the ovarian cycle and the menstrual cycle. The ovarian cycle governs the preparation of endocrine tissues and release of eggs, while the menstrual cycle governs the preparation and maintenance of the uterine lining. These cycles are coordinated over a 22–32 day cycle, with an average length of 28 days.

As with the male, the GnRH from the hypothalamus causes the release of the hormones FSH and LH from the anterior pituitary. In addition, estrogen and progesterone are released from the developing follicles. As with testosterone in males, estrogen is responsible for the secondary sexual characteristics of females. These include breast development, flaring of the hips, and a shorter period for bone growth.

References

- Reproductive-process, reproductive: visiblebody.com, Retrieved 12 May 2020
- 13-3-human-reproduction, biology: opentextbc.ca, Retrieved 25 April 2020
- Process-reproduction-humans-5406051: sciencing.com, Retrieved 18 July 2020
- Sexual-reproduction-in-plants: lumenlearning.com, Retrieved 14 April 2020
- Human-reproductive-system, science: britannica.com, Retrieved 25 May 2020

Permissions

All chapters in this book are published with permission under the Creative Commons Attribution Share Alike License or equivalent. Every chapter published in this book has been scrutinized by our experts. Their significance has been extensively debated. The topics covered herein carry significant information for a comprehensive understanding. They may even be implemented as practical applications or may be referred to as a beginning point for further studies.

We would like to thank the editorial team for lending their expertise to make the book truly unique. They have played a crucial role in the development of this book. Without their invaluable contributions this book wouldn't have been possible. They have made vital efforts to compile up to date information on the varied aspects of this subject to make this book a valuable addition to the collection of many professionals and students.

This book was conceptualized with the vision of imparting up-to-date and integrated information in this field. To ensure the same, a matchless editorial board was set up. Every individual on the board went through rigorous rounds of assessment to prove their worth. After which they invested a large part of their time researching and compiling the most relevant data for our readers.

The editorial board has been involved in producing this book since its inception. They have spent rigorous hours researching and exploring the diverse topics which have resulted in the successful publishing of this book. They have passed on their knowledge of decades through this book. To expedite this challenging task, the publisher supported the team at every step. A small team of assistant editors was also appointed to further simplify the editing procedure and attain best results for the readers.

Apart from the editorial board, the designing team has also invested a significant amount of their time in understanding the subject and creating the most relevant covers. They scrutinized every image to scout for the most suitable representation of the subject and create an appropriate cover for the book.

The publishing team has been an ardent support to the editorial, designing and production team. Their endless efforts to recruit the best for this project, has resulted in the accomplishment of this book. They are a veteran in the field of academics and their pool of knowledge is as vast as their experience in printing. Their expertise and guidance has proved useful at every step. Their uncompromising quality standards have made this book an exceptional effort. Their encouragement from time to time has been an inspiration for everyone.

The publisher and the editorial board hope that this book will prove to be a valuable piece of knowledge for students, practitioners and scholars across the globe.

Index

www.ingramcontent.com/pod-product-compliance
Lightning Source LLC
LaVergne TN
LVHW080713210926
848485LV00003B/461